Light
and plant responses

Light
and plant responses

A study of plant photophysiology and
the natural environment

T.H. Attridge

Senior Lecturer, School of Science, Polytechnic of East London,
England

Edward Arnold
A division of Hodder & Stoughton
LONDON NEW YORK MELBOURNE AUCKLAND

© 1990 T.H. Attridge

First published in Great Britain 1990

Distributed in the USA by Routledge, Chapman and Hall, Inc.
29 West 35th Street, New York, NY 10001

British Library Cataloguing in Publication Data

Attridge, T. H.
  1. Light and Plant responses.
  1. Plants. Effects of day light
  I. Title
  581.19′153

  ISBN 0-7131-2973-5

Typeset in 10/11 pt Times Roman by Mathematical Composition
Setters Limited, Salisbury.
Printed and bound in Great Britain for Edward Arnold, the
educational academic and medical publishing division of Hodder
and Stoughton Limited, Mill Road, Dunton Green, Sevenoaks,
Kent TN13 2XX by Richard Clay, Bungay, Suffolk

Contents

Abbreviations

The abbreviations used in this book are specified here and the first time they are used within the text. Not specified are SI units, which are used wherever possible throughout this book.

Abbreviation	Meaning
ABA	Abscisic acid
A_{meso}/A	mesophyll to leaf area ratio
ATP	adenosine triphosphate
B	blue light
BAP	blue-light-absorbing pigment(s)
B-HIR	blue-light − high-irradiance response
CAM	Crassulacean acid metabolism
Chl-a	chlorophyll-a
Chl-b	chlorophyll-b
Chlide-a	chlorophyllide-a
CFC	chlorofluorohydrocarbon
C-W	Cholodny-Went model
DN	day neutral
DNP	day neutral plant(s)
EPP	phytochrome typical of etiolated material
FR	far-red light
FR-HIR	far-red − high irradiance
GA	gibberellic acid
GLI	gap light index
GPP	phytochrome typical of green tissue
HIR	high irradiance response
IAA	indol-3-acetic acid (auxin)

LAI	leaf area index
LD	long day
LDP	long-day plant
LFR	low fluence response
LHCI	light-harvesting complex I
LHCII	light-harvesting complex II
LHCP	light-harvesting chlorophyll-a/b binding protein
m-RNA	messenger RNA
LVDT	linear variable displacement transducer
n	refractive index
NAD	nicotinamide adenine dinucleotide
NADH	reduced form of NAD
NADP	nicotinamide adenine dinucleotide phosphate
NADPH	reduced form of NADP
PAR	photosynthetic active radiation (400 to 700 nm)
p-Chl	protochlorophyll
p-Chlide	protochlorophyllide
Pfr	far-red-absorbing form of phytochrome
Pfr/Ptot	photoequilibrium of phytochrome
phi	proportion of phytochrome in the Pfr form
PIR	per cent incident radiation
Pr	red-absorbing form of phytochrome
P700	reaction centre of photosynthetic unit
PSI	photosystem I
PSII	photosystem II
R	red light
R : FR	ratio of R to FR
Rubisco	ribulose 1, 5 bisphosphate carboxylase/oxygenase

SD	short day
SDP	short day plant(s)
SOX	low-pressure sodium lamps
UV	ultra-violet light
UV-A	320 to 400 nm
UV-B	280 to 320 nm
VLFR	very low fluence response
WEX	cell wall extensibility
WL	white light
zeta	ratio of R to FR

We trained hard, but every time we were beginning to form up into teams, we would be reorganised. I was to learn in later life that we tend to meet any new situation by reorganising and a wonderful method it can be for creating the illusion of progress while only producing inefficiency and demoralization.

Petronious Arbiter (AD 66)

Preface

Progress in any area of science can be limited by a number of factors, but cannot advance faster than the techniques available to tackle the problems. It would seem that research has two phases to its cycle. In the first a new technique surfaces or, as is often the case in plant sciences, a recent technique is applied to a new discipline. This leads to the elucidation of long-standing problems and often the establishment of new problems which were not previously apparent. This is followed by a phase of fact collection where these techniques are applied more widely, often with greater ingenuity, and the established models and hypotheses are modified or expanded.

The revolution that has occurred in the techniques of molecular biology has greatly encouraged those seeking an understanding of the mechanisms of plant development and this is an unimpeachable pursuit. Such enquiry leads to specialisation and the acquisition of detailed knowledge and it would be possible to lose sight of the way in which the whole plant develops. I thought I might start this book with the quotation 'Lo, though they stand before the pyramids they speak only of the heat of the desert', but after an extensive search of the literature, lasting several minutes, I was unable to make an attribution. Before these new approaches were thrust upon us, research workers tended to be associated with one or another aspect of light-plant interactions, and indeed some areas are so complex it is difficult to see how it could have been otherwise. Here, I have tried to draw together an understanding of how higher plants respond to the natural light environment throughout their life cycle in the hope that in our efforts to understand plant development we can continue the current tendency to take the more integrated approach required by the complex stimuli of the natural situation.

My thanks are due to John Melmoth for his pergamosian endeavours, to Professor Mike Black for reading the chapter on seeds and to my family for their patience during the preparation of this manuscript.

T.H. Attridge
March 1989

1
Introduction

Plants grow as a result of their ability to absorb light energy and convert it into reductive chemical energy, which is used to fix carbon dioxide. Growth enables the species to reproduce its genes, and to do this efficiently the plant must coordinate its development with environmental opportunities.

Plants exist in various associations at a great variety of latitudes and altitudes. They occur under fluence rates which vary daily, seasonally and with shade created by surrounding structures and vegetation. They survive the aquatic environment, the humidity of tropical rain forests and the desiccating effects of the desert. In some environments the stresses upon plants are constant or cyclic, whereas in others they may vary frequently or infrequently. To survive in these habitats plants must retain a high degree of plasticity both as a species and as an individual. Plants must be able to detect the environment and adapt to it.

A significant aspect of a plant's environment is light. As indicated above, a plant has an absolute requirement for light for the purposes of photosynthesis, a process which predominantly requires red (R) and blue (B) wavelengths. If a plant becomes stressed in this regard, it needs to adapt in some way to aid survival. However, these are not the only requirements a plant has of its light environment. Firstly, a plant needs to orientate itself in space so as to present itself to the incident radiation as best it may. Secondly, at higher latitudes in particular, the plant also needs to orientate itself in time so that the various phases of its life cycle are completed within seasons suitable to its habit and ecological niche. Plants need mechanisms capable of detecting a number of variables affecting a large number of processes.

Investigating complex problems such as the relationship between a plant and a varying light environment does not lend itself easily to the scientific process. Although the early workers on photoperiodism were often compelled as a result of the nature of their experiments and available facilities to deal with the variables encountered in the natural environment, the training of the scientist is to ask a single question, design an experiment with a single variable and arrive at an unimpeachable conclusion. The beginning of this painstaking approach to understanding light and plant development has recently been recorded in a booklet entitled *A pigment of the imagination*.[1] This early work is well recorded in almost every text book on plant physiology and will therefore be dealt with only briefly here. It began as the result of the finding of a mutant tobacco plant growing in a field in Maryland, USA. This mutant, which became known as Maryland Mammoth, would continue to grow vegetatively until killed by frost in the winter. Garner and Allard[2,3] artificially shortened the daylength for these plants by bringing them into a windowless shack in the afternoons. After 2 months of this treatment the plants flowered 3 months earlier than plants left under natural light regimes and protected from frost. Simultaneously, these workers were investigating the failure of sequential sowing to produce sequential crops in the soy bean *Glycine max* var. Biloxi. Once again their results showed that daylength was the critical factor controlling flowering. The phenomenon of photoperiodism was discovered, a Pandora's box was opened and some would suggest that hope has not yet emerged.

In the same establishment and not uninfluenced by the presence of Garner and Allard came the work of Borthwick and Parker on the light break of the night period which inhibited flowering of *Glycine max* held under short days. Although this work was originally performed in the late 1930s, it was not published until the early 1950s.[4] The addition of Hendricks to the team led to the elucidation of several action spectra of photosensitive developmental processes. Their results showed that a good number of light-sensitive responses were controlled by a common photoreceptor.[5] Perhaps the most important experiment in the history of the discovery of phytochrome resulted from the collaboration of Eben Toole first with Flint and McAlister and then with the Borthwick team. Flint and McAlister,[6,7] using light-sensitive lettuce seed, had demonstrated not only that maximum stimulation occurred in the R but also that maximum inhibition was found in the far-red (FR). These workers actually used reversibility in their experiments in as much as they raised the germination level of their seeds with R to show the inhibitory capability of FR. It is not possible to judge at this distance in time what significance they attached to this protocol. It was left to Borthwick, Hendricks, Parker, Toole and Toole (1952) to perform the now classical experiment of R/FR reversible lettuce seed germination.[8] Here they showed that the effects of R and FR were not only opposite to one another, they were also antagonistic.

As a result of their experiments the Beltsville workers were able to suggest that plants contain a blue-green pigment to which they gave the rather uninspired name phytochrome (phyto = plant, chrome = pigment). This pigment was the photoreceptor for a number of light-mediated processes and existed in two interconvertible forms. The first form, now known as Pr (red-absorbing-form of phytochrome), was biologically inactive and capable of absorbing R and in so doing was converted to the biologically active form of phytochrome now known as Pfr (far-red-absorbing form). The Pfr form of phytochrome is capable of returning to the Pr form by the absorption of FR. Although the lack of photoreversibility of a response does not disprove the involvement of phytochrome (see very low fluence response (VLFR), p. 28), the demonstration of R/FR reversibility is still regarded as the hallmark of this system.

It was the photoreversibility of phytochrome which led to the development of an *in vivo* assay of the pigment by Butler, Norris, Siegelman and Hendricks.[9] This assay is spectrophotometric and can only be used in etiolated tissue. Chlorophyll absorbs and fluoresces in the same region of the spectrum as phytochrome absorbs. This limitation on the measurement of phytochrome, together with the understandable preference for using monochromatic light instead of broad band irradiation, was to influence research in photomorphogenesis. With a few notable exceptions, experimentation was restricted to the study of germinating seeds and etiolated seedlings. Even today the detection of phytochrome in green tissue is an event limited to a handful of laboratories. Although this approach was to retard our understanding of the role of phytochrome in the natural environment, much was to be learned about the control of seed germination and the biochemical events which occur during the early part of the de-etiolation process. Understanding these changes led to two principal proposals. Mohr suggested that phytochrome acted by repressing and derepressing genes,[10] whereas, Smith hyphothesised that phytochrome acted by controlling membrane permeability.[11] Recent developments in the techniques of molecular biology have added credibility to the involvement of phytochrome with gene expression.[12] Evidence that pytochrome is associated with membranes remains scant despite many distinguished person hours devoted to this problem.

Interest in phytochrome responses of green plants was practically dormant for two decades until a novel approach was provided by the work of Holmes and Smith,[13] which was to be subsequently exploited by Morgan and Smith.[14,15] Phytochrome was revealed to have a role of ecological significance in altering the growth strategies of shade-intolerant species. This work has led others to look at the shade strategies of plants from various habitats.

Phytochrome is not the only pigment involved in developmental processes in plants. Reports that B affects plant development are much older than those involving R. Sachs (1864) showed that the bending of plants towards the light was a process stimulated only by B light.[16] Some believe that this process, now known as phototropism, has received more attention from plant physiologists than a developmental option deserves. Although it can be shown that developing plants are capable of detecting small differences between laterally available light,[17] the majority of plants in the natural environment show no sign of phototropic growth during development. Add to this the fact that the most common plant material used in these experiments is the coleoptile (a tissue which despite its suitability for experimentation is of juvenile and transient importance to the Graminae alone) and the accumulation of this knowledge takes on an inane quality. On the other hand, the coleoptile has provided a model system for understanding the interactions between photoreceptor and hormone, and hormone and cell elongation. Certainly the study of a system so subtle as to lead to an understanding of how the photocontrol of a single hormone can influence both straight and curved growth cannot be without benefit to the understanding of growth in general.

While the involvement of B has long been recognised in phototropic studies, the widespread involvement of B photoreceptor(s) in plant development is a concept which has met with a certain amount of resistance. Whether there is more than one B photoreceptor and by what name it should be called is discussed elsewhere (p. 34). Here it will be referred to as blue-absorbing pigment (BAP), but no bias is implied by the use of the singular form. The purification and isolation of phytochrome was aided by the reversibility of the pigment and put an end to sceptical remarks that phytochrome was 'a pigment of the imagination' or 'a R/FR herring'. No such fortune has been experienced in elucidation of the nature of the BAP. The B region of the spectrum contributes to the energy input of photosynthesis, and both the Pr and the Pfr forms of phytochrome have absorption maxima in the B. The design of early experiments allowed ambiguous interpretation of results, but in recent years techniques have been developed which unequivocally demonstrate a causal relationship between BAP and plant responses. The nature of the BAP still resists elucidation although flavins and carotenoids are perennial favourites (p. 35).

Now that it is accepted that B acting through BAP can mediate photoresponses, a good deal of interest has been directed towards possible interactions between BAP and phytochrome. Obviously, in the natural environment B, R and FR are simultaneously available albeit in various ratios. It is not known if there is any ecological significance to the interaction of BAP and phytochrome.

Some photoresponses have been shown to be controlled by phytochrome, the influence of B having been eliminated. In other photoresponses, the influence of B is thought to have been eliminated but the appropriate fluence range has not been explored; in others the effect of B has not been investigated once phytochrome control has been established. Key to understanding the influence of light and plant development is identifying what is and is not known to influence a response. Although scientific information is seldom complete, there would appear to be more missing from this subject than most. It may be that some of the blame for this situation lies with the investigators, but it also lies with the nature of the scientific literature, which does not lend itself easily to the publication of negative findings.

References for Chapter 1

1   McGee, H. 1987. *A pigment of the imagination*. USDA Publication. Kennon-Kelley Graphic Design.
2   Garner, W.W. and Allard, H.A. 1920. Effects of the relative length of the day and night and other factors of the environment on growth and reproduction in plants. *J. Agric. Res.*, **18**, 553–603.
3   Garner, W.W. and Allard, H.A. 1923. Further studies in photoperiodism, the response of the plant to the relative length of day and night. *J. Agric. Res.*, **23**, 871–920.
4   Borthwick, H.A., Hendricks, S.B. and Parker, M.V. 1952. The reaction controlling floral initiation. *Proc. Nat. Acad. Sci.*, **38**, 929–934.
5   Borthwick, H.A., Hendricks, S.B. and Parker, M.V. 1948. Action spectrum for the photoperiodic control of floral induction of a long day plant Wintex Barley (*Hordeum vulgare*). *Bot. Gaz.*, **110**, 103–108.
6   Flint, L.H. and McAlister, E.D. 1935. Wavelengths of radiation in the visible spectrum inhibiting the germination of light-sensitive lettuce seed. *Smithsonian Misc. Coll.*, **94**, 1.
7   Flint, L.H. and McAlister, E.D. 1937. Wavelengths of radiation in the visible spectrum promoting the germination of light-sensitive lettuce seed. *Smithsonian Misc. Coll.*, **96**, 1.
8   Borthwick, H.A., Hendricks, S.B., Parker, M.V., Toole, E.H. and Toole, V.K. 1952. A reversible photoreaction controlling germination. *Proc. Nat. Acad. Sci.*, **38**, 662–666.
9   Butler, W.L., Norris, K.H., Siegelman, H.W. and Hendricks, S.B. Detection, assay and preliminary purification of the pigment controlling photoresponsiveness of plants. *Proc. Nat. Acad. Sci.*, **45**, 1703–1708.

10   Mohr, H. 1964. The control of plant growth and development by light. *Biol. Rev.*, **39**, 87–112.
11   Smith, H.1972. Phytochrome, hormones and membranes. *Nature*, **236**, 425.
12   Schafer, E., Apel, K., Batschauer, A. and Mosinger, E. 1986. The molecular biology of phytochrome action. In: Kendrick, R.E. and Kronenberg, G.H.M. (eds), *Photomorphogenesis in Plants*. Martinus Nijhoff, Dordrecht, 83–98.
13   Holmes, M.G. and Smith, H. 1975. The function of phytochrome in plants growing in the natural environment. *Nature*, **254**, 512–514.
14   Morgan, D.C. and Smith, H. 1976. Linear relationship between phytochrome photoequilibrium and growth in plants under simulated natural radiation. *Nature*, **262**, 210–211.
15   Morgan, D.C. and Smith, H. 1981. Non-photosynthetic responses to light quality. In: Nobel, P. (ed.), *Encycl. of Plant Physiol.* NS. 12A. Springer-Verlag, Berlin, 109–134.
16   Sachs, J. 1984. Wirkungen des farbigen Lichtsauf flanzen. *Botan. Z.*, **22**, 353–358.
17   MacLeod, K., Brewer, F., Firn, R.D. and Digby, J. 1984. The phototropic responses of *Avena* coleoptiles following localised continuous unilateral illumination. *J. Exp. Bot.*, **35**, 1380–1389.

2

The Natural Light Environment

Light, its nature

Matter, besides having mass, can have electrical charge. Mass acts as a source of a gravitational force field, and charge as a source of an electromagnetic force. The electric field associated with a stationary charge is entirely analogous to the gravitational field associated with a stationary mass. Faraday discovered that moving charges can also produce a magnetic field. Both electric and magnetic fields can accelerate charge. When accelerated, they produce time-varying electromagnetic waves. Maxwell recognized that light is an electromagnetic wave. It has since been realized that not only light but radio waves, infra-red rays, ultra-violet rays, X-rays, and gamma rays are electromagnetic waves. Maxwell's model of light is shown in Figure 2.1. The simple magnetic wave has four properties:

a) speed of propagation C (in a vacuum), $C = 3 \times 10^8 \, \text{ms}^{-1}$ (this is the same for all wavelengths)
b) direction of propagation
c) the wavelength
d) the polarization direction

The direction that E (the orthogonal electric field) points in Figure 2.1 is perpendicular to B (the magnetic field) and the direction of propagation.

At this point the average biologist, having read the above, goes to sit under a tree for about twenty minutes.

Solar radiation

The sun is a star which is approximately 110 times the diameter of the earth and 150×10^6 km distant. The outer layers of the sun consist of approximately 94 per cent hydrogen, 5.9 per cent helium and 0.1 per cent of all other elements. The inner layers are not thought to be dissimilar to the outer layers, and it is from the interior of the sun that solar energy originates. Here, at very high pressures and temperatures, hydrogen is converted to helium. The vast quantities of energy released during this process reach the sun's surface via convection currents and conduction. The production of energy is constant and thus the output of energy is constant, varying only slightly when solar flares and bursts are produced. Light takes approximately 9 min 20 s to travel the

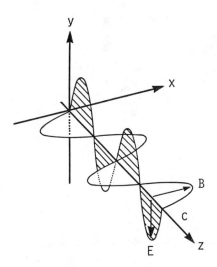

Figure 2.1 Maxwell's model of light.

distance between the sun and the earth. Experimentally determined measurements of the solar spectral distribution agree very closely with those of a black body at 5 800 $^\circ$K. These spectra are shown in Figure 2.2.

It has been calculated that X-rays and gamma rays (2 to 10 nm) and ultra-violet rays (20 to 400 nm) provide 9 per cent of the total solar energy, whereas visible radiation (400 to 800 nm) provides 41 per cent and infra-red (800 nm to 3 000 microns) 50 per cent.

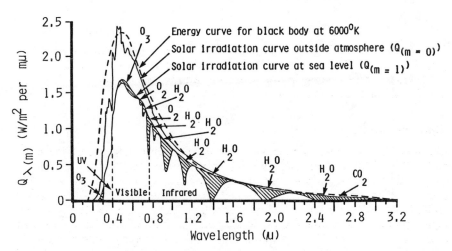

Figure 2.2 Solar spectrum outside the earth's atmosphere compared with that of a black body at 6000 $^\circ$K and with that at sea level (after *Handbook of Geophysics*, revised edition, U.S. Air Force, Macmillan, New York, 1960).

The earth's atmosphere

There is little in space in terms of molecules or particles to alter the spectral energy of sunlight until it reaches the earth's atmosphere. This begins at about 10 000 km from the earth's surface with a layer of hydrogen. This layer has no distinct outer boundary, but at this distance the density of the hydrogen atoms resembles that found in interplanetary space. This said, hydrogen atoms which rotate about the earth's axis can be found as far out as 35 000 km. The light then travels through a layer of helium for approximately 2 400 km, followed by a 900-km layer of atomic oxygen and a thinner layer of about 110 km of molecular nitrogen.

These layers are arranged in order of mass, the heavier gases being closer to the earth's surface. It may be noted that even at the lowest layer (molecular nitrogen) the density of the atmosphere is only about a millionth of that found at sea level. Below these layers, the atmosphere is divided into thermal regions. The nitrogen layer is often described together with the ionosphere as the thermosphere. The high temperatures found at these altitudes are due to the photodissociation and polarization of molecular oxygen by solar radiation. Some of this energy is transferred downwards into the mesosphere. A significant amount of ozone formation takes place in the lower mesosphere, which causes energy absorption at short wavelengths and a concomitant increase in temperature. The majority of ozone formation/destruction takes place in the stratosphere, which accounts for the very high temperatures in the region. The importance of the absorption of short wavelengths to life on earth can scarcely be exaggerated. Its formation prevents higher temperatures in the troposphere and much of the mutagenic effects of ultra-violet light.

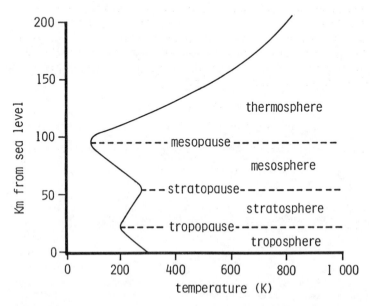

Figure 2.3   The thermal structure of the earth's atmosphere (after Chambers, J.W., 1978).

It is the radiant energy of the troposphere, which is finally absorbed by plants, that is of prime consideration here, and will be dealt with in more detail later. The main source of heat in the troposphere is far-red (FR) re-radiation from the land surface of the earth. In consequence, the troposphere is warmest nearest the earth's surface (Figure 2.3). As a result of industrialisation and the concomitant increase in pollution of the atmosphere with the products of combustive processes, the density of the lower layers of the atmosphere has increased. This has resulted in the so-called 'green-house' effect. The presence of these pollutants in the atmosphere causes absorption of the re-radiated energy which had previously disappeared into space. This is resulting in increases in global temperatures which may cause serious problems.

Atmospheric modification of solar radiation

Solar radiation approaches the earth's atmosphere qualitatively unaltered from the photosphere of the sun. Figure 2.2 shows not only the spectrum of the solar radiation outside the earth's atmosphere but also the spectrum as recorded at sea level. A more detailed spectrum of radiation viewed from the earth's surface is shown in Figure 2.4b. Differences between these spectra are caused by the attenuating effects of the atmosphere. The outer layers are rare and have little effect on incoming radiation. The inner layers of atomic oxygen and molecular nitrogen are, however, capable of absorbing X-rays and gamma rays. These layers coincide with those of the ionosphere. As the atoms or molecules absorb these rays they give up an electron and become positively charged. It is this layer of ions which reflects radiowaves and allows long-distance radio communication. Because the inception of the ionosphere depends upon solar radiation, it tends to weaken on the dark side of the earth and disappear. The ozone layer coincides with the lower mesosphere and upper stratosphere (20 to 55 km above the earth's surface). This layer is vital to life on earth since it absorbs a great deal of ultra-violet light (UV) of less than 320 nm and excludes wavelengths dangerous to life (below 290 nm). The ozone layer is produced by the photolytic cleavage of oxygen by wavelengths of less than 200 nm using nitrogen as an activating partner.

$$O_2 + hv(<200 \text{ nm}) \rightarrow O + O$$

$$O + O_2 + N \rightarrow O_3 + N$$

UV of longer wavelengths (<300 nm) split the ozone in a slower process, viz.:

$$O_3 + hv(<300 \text{ nm}) \rightarrow O_2 + O$$
$$O + O_3 \rightarrow 2O_2$$

The ozone layer is denser in the northern hemisphere (350 D) (Dobson = 0.001 cmbar) than at the equator (245 D). In consequence, UV radiation forms a greater part of the total spectrum in equatorial regions than in northern latitudes. In the late 1970s gaps in the ozone layer were reported over the South Pole. Although the density of the ozone layer was known to vary with factors such as gaseous mixing, solar spot activity and stratospheric temperature changes, the erosion of the ozone layer has now been shown to be due, at least in part, to halogenated hydrocarbons and in particular to chlorinated

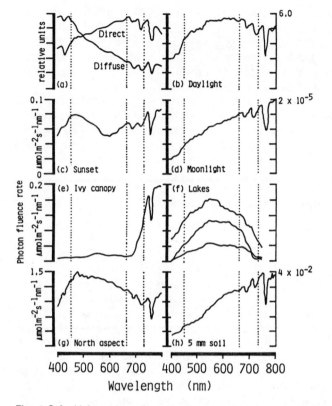

Figure 2.4 Light regimes found in the natural environment (after Smith, H., 1982).
a) Diffuse and direct radiation
b) Clear skies, noon, open situation
c) Sunset
d) Moonlight
e) Ivy canopy-light
f) Lakes
g) North-facing aspect (temperate latitude, northern hemisphere)
h) Soil

fluorocarbons (CFC). These substances are widely used as propellants in aerosol sprays and coolants in refrigeration units and were originally selected for their inert properties. Although long-lived, these compounds finally break down to release chlorine, which interacts with ozone in the following manner:

$$Cl + O_3 \rightarrow ClO + O_2$$

$$ClO + O \rightarrow Cl + O_2$$

Further damage to the ozone layer may be caused by nitrogenous oxides, which are produced by combustive processes:

$$NO + O_3 \rightarrow NO + O_2$$

$$NO_2 + O \rightarrow NO + O_2$$

International negotiations are proceeding to reduce CFC production, but international agreements on environmental issues have always been difficult to achieve in the past. Recent reports suggest that the atmosphere over the North Pole is similarly in danger, and predictions cannot be optimistic. Large increases in UV irradiations over the coming years will have deleterious effects on biological systems.

Rayleigh scattering occurs whenever light meets a particle smaller than its wavelength, and this is a common occurrence as light enters the denser gases of the troposphere. The degree of scattering is inversely proportional to the fourth power of the wavelength. This means that short wavelengths are scattered preferentially, and this gives the characteristic blue appearance to the sky.

Solar radiation received by the world's oceans causes evaporation and thus an atmosphere of considerable humidity. The presence of water droplets and dust particles in the troposphere causes Mei scattering. This phenomenon occurs when light meets a droplet or particle which is larger then the wavelength of the light involved. As the particle size increases (>700 nm) this effect becomes independent of wavelength, producing uniform scattering throughout the visible spectrum. In consequence, regions of the troposphere containing large numbers of water droplets cause a great deal of Mei scattering, most of it away from the earth's surface. These regions, when viewed from the earth's surface appear as dark clouds which are associated with imminent precipitation, whereas white clouds contain fewer water droplets, perform less Mei scattering and inspire poetry.

Finally, water, carbon dioxide and oxygen absorb light of specific wavelengths as a result of their chemical nature. Selective absorption by carbon dioxide occurs at wavelengths beyond the visible and has no effect on the quality of light received by plants. This is not so for water, which has a peak of absorption at 723 nm and other longer wavelengths, whereas oxygen absorbs at 688 and 762 nm.

Under clear skies the above factors account for absorption and reflection of 20 per cent of the total solar radiation reaching the earth's atmosphere. The albedo, or reflective ability of clouds, is a major factor in reducing fluence rates when skies are not clear. Depending on the water content and structure, clouds reflect between 30 and 60 per cent of the incoming irradiation and absorb a further 5 to 20 per cent. The absolute amount of water in the atmosphere depends upon temperature. The higher the temperature, the greater the capacity for the air to carry water. In consequence, at lower latitudes the effect of cloud cover, described above, is at a maximum. The albedo depends upon cloud type and thickness. Stratocumulus, for example, may reflect as much as 80 per cent of the solar radiation it intercepts, but averages around 55 per cent. Most of the reflection disappears into space. With clear skies 80 per cent of the radiation arriving at the outer limit of the atmosphere reaches the earth's surface, but with overcast skies this can be reduced to 20 per cent.

The result of the attenuations described above is that the earth's surface receives skylight or diffuse radiation which is rich in the blue wavelengths, together with sunlight or direct radiation which is relatively rich in the red and far-red region (Figure 2.4a). A combination of these two components results in a spectrum similar to that recorded by Holmes and Smith[1] at noon, on a summer day, at temperate latitude, under clear skies, in an open situation (Figure 2.4b). It may be noted that the fluence rate is fairly uniform between 500 and 700 nm. The fluence rate is lower for shorter wavelengths than for longer wavelengths. Longer wavelengths also show the so-called far-red drop at 762 nm, which is due to attenuation by atmospheric water. It is under a light regime of this quality that the majority of terrestrial plants exist on this planet.

Variations in atmospheric modifications

The main variations are caused by latitude, season, twilight, cloud and dust. These factors can be interrelated.

Latitude and season

The troposphere, if thought of in three dimensions, is an oblate elipsoid. The upper limit, the tropopause, is encountered 16 km above the earth's surface at the equator, at 11 km at lat. 45° and at 8 km at the poles. This, when coupled with the constant 23.5° tilt in the earth's axis, has a number of consequences. The situation prevalent at the winter solstice (named for the northern hemisphere) is as shown in Figure 2.5. Perusal of this figure leads to two main conclusions. At noon on 22nd December, the winter solstice, the sun appears overhead at the tropic of Capricorn. The sun's rays will be passing directly through the atmosphere, and the insolation, that is to say the absorption of

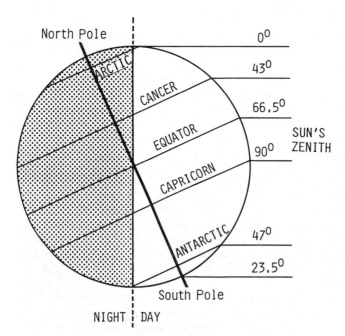

Figure 2.5   The winter solstice (northern hemisphere) showing the angle of the sun's zenith.

radiant energy by a flat surface, will be at its greatest. In other words, the sun's energy will be spread over the smallest area at this time of year, at this latitude. As latitude increases from the tropic of Capricorn to the South Pole, and from the Equator to the North Pole, the solar zenith decreases and insolation decreases. Daylength, however, has the opposite correlation. On the winter solstice the daylength is 12 h at lat. 23.5°S and 24 h at lat. 66.5°S. Thus at low latitudes, solar radiation passes through less atmosphere and strikes the ground at a near vertical angle but for a comparatively short period. At more southerly latitudes the sun's rays pass through more atmosphere, exaggerating the attenuating process described above, and fall on to a larger area but for a longer period of time. The effect of the increased path length through the atmosphere at these latitudes is slightly reduced since the troposphere is considerably thinner here than at lower latitudes.

North of the tropic of Capricorn on the winter solstice, the solar zenith decreases with latitude to the Equator and then decreases with increasing latitude in the northern hemisphere. Daylength also decreases in this direction until at the Arctic Circle the sun cannot be seen above the horizon. These effects are of course seasonal. By the spring equinox, March 21st, the earth has completed a further quarter of its orbit, the zenith at the Equator is 90°, at the tropics 66.5°, and zero at both poles. The daylength is the same all over the earth. By the summer solstice, the earth has completed another quarter of its orbit and essentially the opposite situation of that described for the winter solstice is operative. The autumnal equinox, September 22nd, is of course identical to spring equinox.

Latitude and twilight

Twilight has received several definitions but here it will be regarded as the periods of time when the sun is within $+10°$ and $-10°$ of the horizon and refers to light quality at dawn and dusk.

Figure 2.4a shows that in an open situation the daylight spectrum can be divided into two components, direct and diffuse radiation. When the sun is low in the sky, direct radiation passes through more atmosphere than when the sun is at its zenith. Thus the attenuating effects are greater and consequently diffuse radiation makes a larger contribution to the spectrum received on the earth's surface, as shown in the sunset spectrum (Figure 2.4c). When the solar disc is below the horizon obviously the low radiation level available is entirely diffuse radiation. This radiation is at its greatest attenuation since it is being reflected from dust and clouds. The path length of the radiation is such that light effectively passes through the troposphere twice. Spectra are not available for below horizon twilight, but the sunset spectra of Holmes and Smith[1] show relative increases in the blue and, according to these authors but not according to Hughes *et al.*,[2] a decrease in the ratio of R : FR (for definition see p. 16) during sunset. The absence of a decrease in R : FR in the work of Hughes *et al.* was explained in terms of clouds on the horizon. Water vapour absorbs FR whereas the decrease in R : FR reported previously[1] is explained in terms of the longer path length of the light through the ozone layer. Ozone absorbs in the red-green region of the spectrum, resulting in a relative increase in FR.

The duration of twilight depends upon the thickness of the atmosphere and

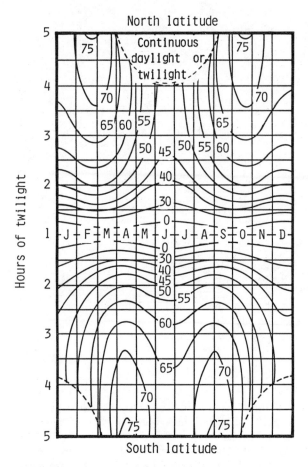

Figure 2.6   Duration of twilight on a world basis (from *Physical Geography*. Strahler; A.N., Wiley and Sons, New York-London 1969).

the rate at which the sun sinks below the horizon. Both factors vary with latitude but in opposite directions. The atmosphere, in particular the troposphere where refraction, reflection and attenuation take place, is thicker at the Equator than it is at the poles. This would lengthen the period of twilight since the increased path length would enable the sun to sink to a lower level before the refracted and reflected rays were occluded. However, the prime factor in twilight length depends on the apparent path of the sun in the sky, which in turn depends upon latitude.

Figure 2.6 shows the twilight situation on a world-wide basis on the summer solstice. At the Equator twilight has a duration approximately 1.25 h, whereas from latitude 50°N northwards continuous daylight or twilight with a low solar elevation is enjoyed. On the same day at 50°S the twilight lasts 2.75 h. The opposite situation for 50°S would exist for the winter solstice.

Aspects of aspect

The effect of aspect varies with topography, time of day, latitude and season. A north-facing slope, in the northern hemisphere, in the summer under clear skies at noon receives light which is composed largely of diffuse radiation (Figure 2.4 g) rich in blue wavelengths but lacking longer wavelengths. Smith[3] gives a R : FR for this situation of 1.3, which is higher than the majority of readings given for noon, clear skies, open situation, but the difference is not known to be of any significance to plant development. However, the total fluence rate is 25 per cent of the open situation and this is the most significant aspect of aspect. The decrease in total fluence in this situation is greater the higher the surrounding topography. The decrease is also more marked with latitude and will be greater in the winter than in the summer and at times when the sun is low in the sky. Similar parameters control the small-scale effects of neutral density shade provided by tree trunks and branches, walls, rocks and buildings.

Holland and Steyn[4] present a model of the annual short-wave energy load of slopes of different angles, aspects and latitude. They predict that at lat. 45° N/S, the compass orientation of slopes will cause the greatest variation in local vegetation. This effect decreases towards the poles or the Equator. A literature review of vegetation variations in a large number of geographical locations would seem to support this view.

Global energy

Diurnal absorption of energy, or insolation, by the flat surface of the earth depends upon a number of factors. The major factors, daylength and solar elevation, have been previously discussed. Global models for energy absorption by the earth's surface[5] show, as might be supposed, that the energy absorbed near the Equator is nearly constant throughout the year, whereas at higher latitudes large seasonal fluctuations are experienced. This and similar models ignore the variation in total energy absorption caused by sea and land, by vegetation and desert, and by differing cloud structure and cover.

Leaf and light interactions

As has been mentioned in previous sections, the light received by a leaf surface varies with latitude, season, time of day, aspect, leaf inclination and cloud cover. Part of this radiant energy is reflected, part absorbed and part transmitted. These effects are quantified in Figure 2.7. The albedo of the leaf surface is approximately 20 per cent of the total incident radiation and consists largely of green light reflected from the chloroplast and FR reflected from air-water interfaces between cell walls and intercellular spaces. The spongy mesophyll of dicotyledons is particularly efficient in FR reflection. It is the FR region of the spectrum which is generally used for the remote sensing of vegetation by satellites. It was hoped to refine this technique to quantify vegetation and even identify some species. However, partly as a consequence of FR absorption by cloud cover, these hopes have foundered. An examination of the absorption spectra of the most common chloroplast pigments, viz.

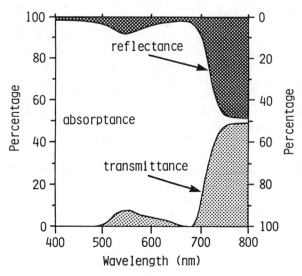

Figure 2.7   The reflectance, absorptance and transmittance of a green leaf (after Smith, 1986).

chlorophyll-a and -b, and carotenoids (see Figure 3.2) reveals why light passing through a green leaf or canopy of leaves would be progressively depleted of B (400 to 500 nm) and R (600 to 700 nm). Light transmitted from a leaf is relatively rich in yellow-green light (500 to 600 nm) and the FR (700 to 800 nm).

Canopy light quality

Not all light beneath a canopy has been transmitted through leaves. The relative importance of transmitted light to canopy light varies with the canopy structure, depth, density and species.[6] Few canopies are entirely closed, and through gaps in the canopies arrives a contribution of both diffuse and direct radiation. To this is added green light via leaf transmission and FR via transmission and reflection from reflexed leaves. Figure 2.4e shows a spectrum taken beneath a canopy of ivy leaves. Fluence is reduced across the spectrum but is relatively rich in the FR region. There is also a slight relative enrichment in the green, which is more marked in other species.

Plants growing beneath a canopy have adapted in various ways to deal with a light environment which has reduced photosynthetically active radiation (PAR), i.e. 400 to 700 nm, and low red/far-red ratio (R : FR). R : FR is referred to in older texts by the Greek letter zeta (ζ), has been measured in various ways, but Smith[3] uses the following definition:

$$R : FR = \frac{\text{photon fluence rate in 10 nm band centred on 660 nm}}{\text{photon fluence rate in 10 nm band centred on 730 nm}}$$

Smith also presents a selection of R : RF found in the natural environment. These findings are reproduced in Table 2.1.

Table 2.1 Total photons (400–800 nm) and R : FR in natural environment.

Situation	μmol m^{-2}s^{-1}	R : FR
Daylight	1900	1.19
Sunset	26.5	0.96
Moonlight	0.005	0.94
North Aspect	480	1.3
Ivy canopy	17.7	0.13
Lakes—1 m depth		
Black loch	680	17.2
Loch Leven	300	3.1
Loch Borralie	1200	1.2
Soil (5 mm)	8.6	0.88

(After Smith 1982). The above data were derived from the spectroradiometer scans which provided Fig. 2.4 and are not to be regarded as environmental means.

Gap light index (GLI)

When gaps arise in the canopy, a change occurs not only in R : FR but also in PAR, and it is this increase in available resources which allows an increase in the plant biomass beneath the gap. Having said this, it is important to realise that while competition for light has been temporarily removed, the availability of water and minerals in the soil of the forest floor may become critical.[7] The GLI has been developed[8] to enable the prediction of the percentage incident PAR at any point in or around the gap. These predictions are made from the geometry of the gap in the canopy.

The GLI is of the form

$$\text{GLI} = [(T_{\text{diffuse}}\ P_{\text{diffuse}}) + (T_{\text{beam}}\ P_{\text{beam}})] \cdot 100.0$$

where P_{diffuse} and P_{beam} are the proportions of the incident seasonal PAR received at the top of the canopy as either diffuse skylight or direct radiation respectively. P_{diffuse} is derived, $P_{\text{diffuse}} = 1 - P_{\text{beam}}$. The values T_{diffuse} and T_{beam} are the proportions of diffuse and direct radiation that are transmitted through the gap to a point below. Extensive advice on how to obtain or derive these values is offered.[8]

The index ranges from 0, where there is no clear gap in the canopy, to 100, which represents the open situation. The geometry of the gap for this model was described in terms of its spherical coordinates—in other words, the angle from vertical to the edge of the gap in each compass direction. Thus for any gap investigated, a set of gap coordinates can be derived for any point of interest in the understorey.

This model was tested in a mixed woodland using PAR measurements in two gaps at 30 points subjectively chosen for their variation. The relationship between percentage transmission and GLI derived from these data shows a reasonable straight line relationship with a slope not significantly different from 1. Using this index with an understanding of the way in which it is affected by factors such as slope, aspect and season will enable better comparisons of the ecology beneath canopy gaps.

Sunflecks

The qualification and quantification of light in sunflecks is fraught with practical difficulties due to their transient and mobile natures. A sunfleck is predominantly composed of direct radiation at its inception as it breaks through the canopy.[13] The lifetime of a sunfleck varies with canopy; ivy produces a sunfleck with a half life of 0.01 to 15 s,[9] whereas large gaps in woodland canopies produce sunflecks lasting 20 min.[10] Sunflecks of this duration may have consequences for plant development beneath canopies (see Chapter 5). Holmes and Smith[1] show that sunflecks in wheat canopies decrease in fluence rate down through the crop but produce only a small decrease in R:FR, whereas subsequent work[11] on woodland sunflecks shows that the downward reflection of leaves leads to non-homogeneous sunflecks with areas of both high and low R:FR. It is the areas of high R:FR that are of consequence to the plants beneath the canopy since the areas of low R:FR are qualitatively similar to the surrounding canopy light.

Soil and light quality

It is clear from the work of Wooley and Stroller[12] that the transmission of light through soil is poor. A 2-mm layer of sand transmits less than 2 per cent of the incident radiation, i.e. approximately 40 μmol m^{-2}s^{-1} at noon, at temperate latitude, clear skies, open situation. Most of the radiation is between 700 and 800 nm.

Silty clay loams are completely opaque at a depth of 2 mm but allow transmission throughout the spectrum to shallower depths. The presence of moisture in sand decreases the number and sharpness of discontinuities in refractive indices, which increases transmission. In loam, water reduces reflection between grains since it dissolves and suspends dark material and causes a decrease in transmission. Sand has a greater effect on R:FR than loam does. Frankland[13] shows that daylight passing through 1 mm of silty loam, moist sandy soil and dry sandy soil has R:FR of 1.0, 0.6 and 0.4 respectively. These findings were confirmed by Bliss and Smith,[14] who went on to investigate penetration of leaf litter by light.

Litter and light quality

Very little attention has been given to the attenuating effects of leaf litter on light received by soil. However, Bliss and Smith[14] show that light transmitted through 15 mm of Corsican Pine needles (*Pinus nigramaritima*) is absorbed uniformly across the spectrum with great efficiency. Only 0.5 to 1.0 per cent of incident radiation is transmitted and this is thought to occur through the spaces between the needles. By contrast, fresh oak leaves (*Quercus robur*) produce a transmission spectrum not atypical of heavy shade with a relative increase in longer wavelengths which produces a R:FR of Ca 0.3.

Aquatic light

The penetration of light into water depends initially on the solar elevation. Low elevations cause 50 per cent reflection, which decreases rapidly to 5 per cent or less when the sun is less than $10°$ above the horizon. Furthermore, this reflection is dependent on cloud cover and wind speed. Light which penetrates the water surface is subjected to Rayleigh scattering and is absorbed in the FR region by water molecules. In consequence, B and FR penetrate water poorly; hence the under-water spectrum can become relatively rich in the red region. Some lakes (Crater Lake, USA) contain so little dissolved material that the absorptions and attenuations resemble those of distilled water, the decrease in energy being logarithmic in accordance with Beer's Law.[15] In such water R : FR will increase linearly with depth and at any depth will be dependent on the incident R : FR. This is the only situation in the natural environment where R : FR can rise much above unity. However, the absolute fluence of red is low, and in the majority of lakes which have an algal population, this radiation is rapidly absorbed. Spence[15] points out that the majority of angiosperms do not grow at depths exceeding 6 m and only rarely in those exceeding 4 m, although representatives of the Bryophyta and Chlorophyta have been reported at greater depth. The spectra of three fresh water lakes which vary in PAR due to turbidity are shown in Figure 2.4f. The light profiles of lakes vary with season generally as a result of algal bloom. Under water the R : FR increases as a result of FR attenuation by water.

Chambers and Spence[16] discuss the changes which take place in the underwater environment at dusk in three Scottish lakes. They point out that in natural water with a significant alga population the R : FR of the downwell (radiation measured with the spectroradiometer head pointed vertically upwards) is dependent not only upon the attenuation coefficient of water but also on those of the algae, water-soluble, light-absorbing substances and the incident R : FR. When attenuation by algae is high and that of dissolved substances low, the R : FR at any given depth can be determined by the chlorophyll-a content of the algae in the water and can lead to decreases in the R : FR. The decreases in R : FR which take place at twilight as a result of atmospheric attenuation will, of course, alter the R : FR of the incident radiation and exaggerate the decrease in the ratio during these periods of the day. The variation in the underwater R : FR between twilight and midday is implicated in a number of light-controlled developmental processes in aquatic angiosperms (see p. 131)

Maritime waters vary in turbidity quite markedly. In filtered sea water about 40 per cent of the incident radiation reaches 1 m in depth and 22 per cent reaches 10 m. Average turbid sea water has corresponding readings of 35 and 10 per cent, whereas turbid coastal water has 23 and 0.5 per cent. It has been suggested[17,18] that there are three factors which determine turbidity in sea water: particulate detritus, dissolved organic compounds and phytoplankton. Only phytoplankton absorbs R, whereas all absorb B. The clearest ocean water is found in the Sargasso Sea in the central north Pacific. Here 1 per cent incident radiation can be measured at 250 m, compared to coastal waters where 1 per cent incident radiation is measured at 8 m. The generalisation is that coastal waters are turbid and this decreases as the water becomes deeper.

Moonlight and starlight

Moonlight is reflected solar radiation and becomes the main source of radiation when the solar disc is more than 10° below the horizon. The reflective ability, or albedo, of the moon's surface is poor, often quoted as 7 per cent.[19] The surface of the moon produces Mei scattering and refraction in the reflected light, which is then subjected to the usual attenuations by the earth's atmosphere. As a result of these factors, moonlight viewed from the earth's surface produces a spectrum like that shown in Figure 2.4d. This spectrum is low in total fluence and relatively rich in the longer wavelengths; a R:FR of 0.94 has been supplied.[20] Global rather than individual starlight is not dissimilar to moonlight, but in quantitative terms it is much lower in fluence rate.

Many research workers in plant physiology have extrapolated from periods of artificial darkness to night length in the natural environment, and it is therefore important to discover whether the low fluence rates available at night are capable of influencing plant processes. It may be seen from the measurements of *in vivo* phytochrome photoconversions in etiolated tissue[21] that the energy supplied by moonlight and starlight is capable of converting Pr to Pfr, but it has also been argued that the attenuating effects of chlorophyll would prevent this photoconversion. Wooley[22] shows that when light passes through a green leaf the R fluence reaching the lower epidermis is reduced by an order of magnitude. This might be thought to preclude phytochrome action, but

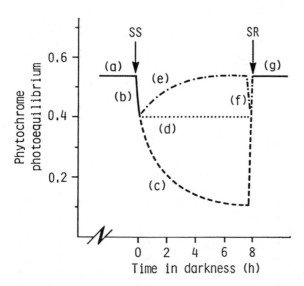

Figure 2.8 Hypothetical changes in phytochrome photoequilibrium during the night versus the dark period: (a) represents the Pfr level in daylight which would rapidly fall (b) under the lower R:FR experienced at sunset when horizons are clear. On a moonless or overcast night Pfr would continue to revert (c) until it reaches a low level. Moonlight might allow the Pfr level to approach that found in daylight (e). The drop (f) is due to the low R:FR experienced early during sunrise. SS = sunset, SR = sunrise (after Holmes and Wagner, 1981).

Holmes and Wagner[23] point out that phytochrome-mediate responses can be initiated at light levels two orders of magnitude below this level. It would seem from theoretical predictions, for dicotyledons at least, that phytochrome would behave as shown in Figure 2.8. The level (a) would represent the Pfr level in daylight, which would rapidly fall (b) under the lower R:FR experienced at sunset when horizons are clear. The fate of Pfr then depends on the phase of the moon and cloud cover. On a moonless or overcast night Pfr would continue to revert (c) until it reaches a low level. On the other hand, there might be enough energy in moonlight to maintain the Pfr remaining after sunset to nearly regain the level of Pfr found in daylight (e). The drop (f) is due to the low R:FR experienced early during sunrise. Consideration of moonlight as light of physiological importance needs to take into account not only the R energy of the spectrum but also the antagonistic effects of FR,[24] the presence or absence of reversion and the effect of lower night temperatures on this system. Experiments continue in this interesting area.

References to Chapter 2

1   Holmes, M.G. and Smith, H. 1977. The function of phytochrome in the natural environment. I. Characterisation of daylight for studies in photomorphogenesis and photoperiodism. *Photochem. Photobiol.*, **25**, 533–538.

2   Hughes, J.E., Morgan, D.C., Lamberton, P.A., Black, C.R. and Smith, H. 1984. Photoperiodic time signals during twilight. *Plant Cell Environ.* **7**, 269–277.

3   Smith, H. 1982. Light quality, photoperception and plant strategy. *Ann. Rev. Plant Physiol.*, **33**, 481–518.

4   Holland, P.G. and Steyn, D.G. 1975. Vegetational responses to latitudinal variations in slope angle and aspect. *J. Biogeog.*, **2**, 179–183.

5   Henderson, S.T. 1977. *Daylight and Its Spectrum*. Hilger, Bristol, 39.

6   Goodfellow, W. and Barkham, J.P. 1974. Spectral transmission curves for a Beech (*Fagus sylvestris*) canopy. *Acta Bot. Neerl.*, **23**, 225–230.

7   Vitousek, P.M. and Denslow, J.S. 1986. Nitrogen and phosphorous availability in treefall gaps of a lowland tropical rainforest. *J. Ecol.*, **74**, 1167–1178.

8   Canham, C.D. 1988. An index for understorey light levels in and around canopy gaps. *Ecology*, **69**, 1634–1638.

9   Norman J.M. and Tanner C.B. 1969. Transient light measurements in plant canopies. *Agron. J.*, **61**, 847–849.

10 Woods, D.B. and Turner, N.C. 1971. Stomatal response to changing light by four species of trees varying in shade tolerance. *New Phytol.*, **70**, 77–84.

11 Tasker, R. 1977. Ph.D. thesis, Nottingham University.

12 Woolley, J.T. and Stroller, E.W. 1978. Light penetration and light induced seed germination in soil. *Plant Physiol.*, **61**, 597–600.

13 Frankland, B. 1981. Germination in the shade. In: Smith, H. (ed.), *Plants and the Daylight Spectrum*. Academic Press, London, 187–204.

14 Bliss, D. and Smith, H. 1985. Penetration of light through soil and its role in the control of seed germination. *Plant Cell Environ.*, **8**, 475–483.

15 Spence, D.H.N. 1981. Light quality and plant responses underwater. In: Smith, H. (ed.), *Plants and the Daylight Spectrum*. Academic Press, London, 245–275.

16 Chambers, P.A. and Spence, D.H.N. 1984. Diurnal changes in the ratio of underwater red to far-red light in relation to aquatic plant photoperiodism. *J. Ecol.*, **72**, 495–503.

17   Jerlov, N.G. 1966 Aspects of light measurement in the sea. In: Bainbridge, R., Evans, G.C. and Rackham, O. (eds), *Light as an Ecological Factor*. Blackwell, Oxford, 91–98.

18   Jerlov, N.G. 1966. Light measurements in the sea in terms of quanta. In: Evans, G.C., Bainbridge, R. and Rackham, O. (eds), *Light as an Ecological Factor II*. Blackwell, Oxford, 521–524.

19   Kopal, Z. 1969. Photometry of scattered moonlight. In: Kopal, Z. (ed.), *The Moon*. Holland Reide, Dordrecht, 357–370.

20   Salisbury, F.B. 1981. Twilight effect: initiating dark measurement in photoperiodism of *Xanthium*. *Plant Physiol.*, **67**, 1230–1238.

21   Furuya, M. and Hillman, W.S. 1964. Observations on spectrophotometrically assayable phytochrome *in vivo* in etiolated *Pisum* seedlings. *Planta.*, **63**, 31–42.

22   Wooley, J.T. 1971. Reflectance and transmittance of light by leaves. *Plant Physiol.*, **47**, 656–662.

23   Holmes, M.G. and Wagner, E. 1980. A re-evaluation of phytochrome involvement in time measurement in plants. *J. Theor. Biol.*, **83**, 255–265.

24   Smith, H. 1983. The natural radiation environment: limitations on the biology of photoreceptors. Phytochrome as a case study. In: Cosens, D.J. and Vince-Prue, D. (eds), *The Biology of Photoreception*. Society of Experimental Biology Symposium XXXVI, 1–18.

3
Light-absorbing pigments

Introduction

Plants contain a number of light-absorbing pigments which vary in both chemical composition and function. All the functions of all the pigments are not fully explained. For example, the presence of anthocyanins in flower parts can be explained in terms of their absorption of ultra-violet (UV) wavelengths and the consequent attractiveness to insects, whereas the function of anthocyanin and related pigments when it appears in other parts is less readily understood. In this chapter we consider the main plant pigments to which roles can be assigned. These include the photosynthetic pigments, cholorophyll and carotenoid, and the developmental pigments, phytochrome and BAP (blue-light-absorbing pigment). Phototropism is also considered here, rather than in following chapters, since it is almost entirely under the influence of BAP and is not often part of vegetative development in the natural environment.

Photosynthetic pigments

Chlorophyll

The prime pigment of photosynthesis common to all oxygen-evolving plants is chlorophyll-a (Chl-a). This molecule has two constituents. The first, often referred to as the 'head' of the molecule, is a cyclic tetrapyrrole complexed with a central magnesium ion; the second, the 'tail' of the molecule, is a phytol side chain. Light absorption is a function of the 'head' of the molecule, the 'tail' playing no part in this process. The structure of Chl-a is shown in Figure 3.1 and the absorption spectrum in Figure 3.2. Higher plants also contain Chl-b, which differs from Chl-a in that a -CH_3 group on ring II is replaced by a -CHO group. The inclusion of this group leads to a significantly different absorption spectrum (see Figure 3.2).

Carotenoids

Higher plants also contain carotenoids. These yellow-orange pigments are long-chained tetraterpenes and in consequence lipophilic. About 500 different compounds have been identified in the plant kingdom and these can be divided into two groups, the carotenes, which are oxygen-free, and the oxygen-containing xanthophylls. In higher plants the common form of carotenoid is

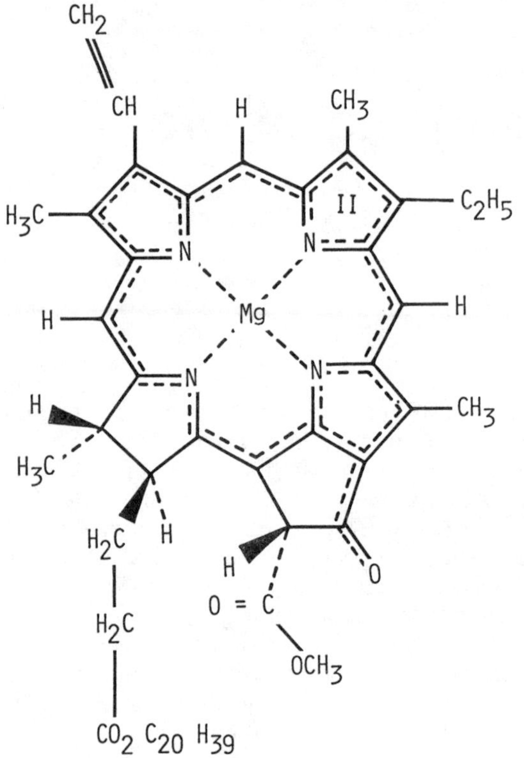

Figure 3.1   Structure of chlorophyll-a. The $-CH^3$ group on ring II is replaced by a $-CHO$ group in Chl-b (after Hipkins, 1984).

β-carotene. This is a symmetrical molecule with 11 conjugated bonds all in the *trans* position. Xanthophylls contain oxygen in various groups, viz. hydroxy, carboxy, methoxy and epoxy. The most common form of xanthophyll in higher plants is lutein, which is located in the leaves and contains two hydroxy groups. The yellow appearance of petals and fruits is also a result of the presence of xanthophyll. Carotenoids absorb B and usually have a three-peaked spectrum between 400 and 500 nm. This broad absorption spectrum (see Figure 3.2) is of advantage to the plant since carotenoids function as accessory pigments for photosynthesis. This is to say, they form part of the light-harvesting apparatus of the photosynthetic unit, transferring the absorbed energy with about 40 per cent probability. A second function of carotenoids is to protect chlorophyll from photo-oxidation under conditions of high fluence rate. Chlorophyll acts as a photosensitizer, producing singlet oxygen which carotenoids are capable of inactivating. BAP has not been identified but carotenoids, together with flavins, are likely candidates, the main line of evidence being the similarity between absorption spectra of these compounds and action spectra of developmental processes. This is discussed in more detail later (p. 35).

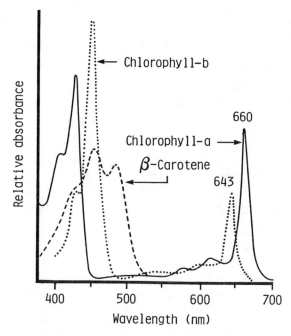

Figure 3.2 Absorption spectra of chlorophyll-a, -b, and -carotene (redrawn from Halliwell, 1984, and Hipkins, 1984).

Phytochrome, its nature

The pigment phytochrome is a biliprotein and as such consists of two major components. The first is the bilin, which is the chromophore of phytochrome, that is to say the part of the molecule which absorbs light and causes a photochemical reaction. The chromophore is shown in Figure 3.3. Phytochrome is a photochromic pigment, which means it exists in two forms, and the photochemical reaction induced by the absorption of light converts phytochrome from one form to the other. The two forms of the pigment are known as Pr and Pfr and the absorption spectra of both forms are shown in Figure 3.4. The names of these forms derive from their absorption maxima.

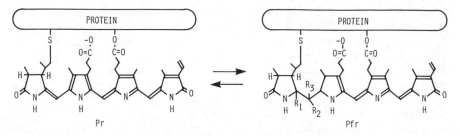

Figure 3.3 Chromophore of phytochrome (redrawn from Pill-Soon Song, 1984).

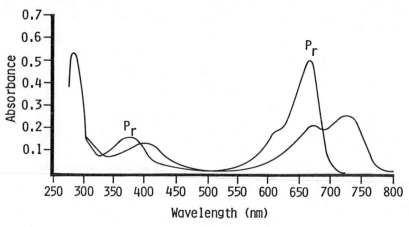

Figure 3.4   The absorption spectra of Pr and Pfr purified from rye (after Smith and Daniels, 1981).

Initially, this convention does not strike everyone as totally logical because when Pr does absorb R it is rapidly converted to Pfr and, conversely, when Pfr absorbs FR it is converted to Pr.

$$Pr \underset{FR}{\overset{R}{\rightleftharpoons}} Pfr$$

The conversions of Pr to Pfr and *vice versa* are not direct events but pass through a number of intermediates. These intermediates exist for only a very short period of time and have been measured by a variety of techniques generally involving spectrophotometry at low temperatures. The convention by which they are named is now less logical than it used to be. The appearance of the intermediates may require light, in which case they receive the prefix 'lumi-'; if however they can be produced from their precursor in darkness, they receive the prefix 'meta-'. If the intermediate originates from Pr it is labelled R and if from Pfr it is labelled F. Finally, the intermediates receive alphabetical suffixes to denote the order in which they were originally thought to arise. The logic of this system has now been disturbed by the current idea that meta-Rb is not on the main pathway from Pr to Pfr or ubiquitous in occurrence but is formed by light from meta-Rc to which it is capable of returning in darkness. The intermediates are shown in Figure 3.5. Most of the intermediates have very short half lives and have yet to be shown to be of any physiological significance. However, it has been shown that dehydration can inhibit the formation of some of these intermediates.[1] This is of significance to seed germination. During ripening on the parent plant there must come a stage when sufficient water has been lost to prevent Pr/Pfr interactions and the ratio of these two forms becomes fixed largely under the influence of the pre-dominating light regime. This can have significance for the light requirements of seeds after rehydration (p. 54). In early literature references to a process called inverse dark reversion can be found. Pfr was known to revert to Pr in the dark but inverse dark reversion sought to explain the appearance of Pfr

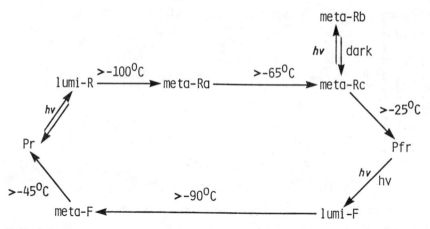

Figure 3.5 Phytochrome intermediates (redrawn from Eilfeld and Rudiger, 1985).

under these conditions. It is now understood that a limited amount of Pfr can be derived from the intermediate meta-Fa in darkness. It may well be that under natural daylight the intermediates between Pr and Pfr and *vice versa* function to preserve the level of phytochrome.[2]

The exact changes of the phytochrome molecule between Pr and Pfr and between the various intermediates is not fully understood. These problems are beyond the scope of this book and are fully explored elsewhere.[3]

The second part of the phytochrome molecule is the protein. After a number of artifacts in the extraction procedure have been rectified the molecular weight of native or intact phytochrome is thought to vary between 120 and 127 kDa according to species. The sequence of the amino acids of the protein is now known together with much of the secondary and quaternary structure of phytochrome. This subject has been extensively reviewed by Viestra and Quail (1986)[4], who point out that although a number of hypotheses about the mode of action of phytochrome suggest this pigment is bound to a membrane, the hydropathic properties of the protein are more characteristic of a soluble globular protein rather than an intrinsic membrane protein.

Pfr, the active form of phytochrome

Since the early days of research into phytochrome-stimulated responses, most, although not all, experimental findings are in keeping with the view that Pfr is the active form of phytochrome. That is to say, the presence of Pfr at a particular level stimulates the expression of the response. Small alterations in the availability of Pfr have been shown to affect phytochrome-controlled systems. On the other hand, Smith (1983)[5] suggests that Pr, Pfr/Pr or Pfr/P could also be the stimulatory entity. Whereas these unresolved problems are of importance to those concerned with elucidating the primary action of phytochrome, the correlation of Pfr levels with phytochrome responses colours the understanding of phytochrome responses. Once produced, the fate of Pfr may be various. It might be photorevesed to Pr, or it may be consumed in some way while interacting with the metabolism of the cell, and a high proportion might be destroyed, a process referred to as destruction or decay. *In vivo* destruction, which has been studied for many years,[6] was once thought to be important to the mode of action of phytochrome,[7] but is now more fully understood.[8] Lastly, Pfr may be subject to thermal (dark) reversion to the Pr form.[6] Little is known of this process other than it must be different, at least in part, from the FR-stimulated photoreversion from Pfr to Pr.

Low fluence response (LFR) and the very low fluence response (VLFR)

One of the hallmarks of the phytochrome system is that light treatments of low energies are required to convert Pr to Pfr and *vice versa*. LFR is characterised as requiring energy in the range of 1 to 1 000 $\mu mol\,m^{-2}$ and is exemplified by responses such as the germination of Grand Rapids, a light-sensitive variety of lettuce, which could be stimulated by short periods of R and reversed by short periods of FR. Repeated treatments of R and FR can be given and the germination response always depends upon the last irradiation.[9] In more recent years phytochrome responses sensitive to energies in the range of 10^{-1} to 10^{-4} $\mu mol\,m^{-2}$ have been revealed and dubbed VLFR. These responses include rapid chlorophyll synthesis[10] and have been studied particularly closely in the development of coleoptiles in cereals,[11] where a role can be discerned for them in the natural environment (see p. 67). The relationship between the VLFR and LFR and the appropriate energy level is shown in Figure 3.6. It may be noted that these VLFR are produced by such low levels of Pfr that they can be produced by FR irradiations and are certainly not reversible by light regimes found in the natural environment.

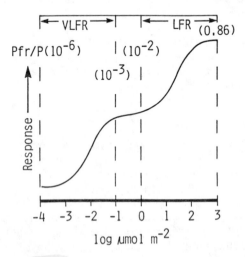

Figure 3.6 Fluence range of the low fluence response (LFR) and the very low fluence response (VLFR) of phytochrome. The proportion of total phytochrome as Pfr also shown (from Kronenberg and Kendrick, 1986).

High irradiance response (HIR)

The HIR is not normally present in fully de-etiolated tissue but has been studied extensively in seeds (p. 51) and etiolated plants. The relevance of the HIR to the natural situation is therefore thought to be restricted to the species in which it affects germination and the emergence of etiolated seedlings from the soil. Hartmann,[12,13] in his classical work on the action spectrum of lettuce hypocotyl growth (Figure 3.7), showed that with long-term irradiation a maxima of inhibition of elongation could be obtained at 717 nm. By using dichromatic irradiations he showed that this HIR was mediated by phytochrome and dependent on the existence of a narrow range of Pfr/Ptot. Etiolated *Lactuca sativa* var. Grand Rapids were simultaneously irradiated with 768 nm and 658 nm. A constant fluence rate of FR was given while the R fluence rate was varied. Neither of these wavelengths are effective in causing inhibition when used separately, but when used together, at certain fluence rate ratios, optimal inhibition of hypocotyl elongation is obtained. From this type of investigation it was deduced that maximum inhibition of hypocotyl elongation could be attained when the Pfr/Ptot ratio was held at 0.03. When the Pfr/Ptot was held above 0.3 or below 0.002 the inhibition of hypocotyl elongation was lost. HIR responses in general are fluence-rate-dependent and it has been suggested by several authors that this relationship is due to the cycling rate of phytochrome. The more energy absorbed by the phytochrome system the faster one form converts to the other and this, it is suggested, affects the action of phytochrome in some way. Hartmann's work was the first to establish the FR-HIR. B is also thought to contribute to a B-HIR. However, the responses are complicated to interpret since B may act via BAP or B − HIR, or interaction may occur between phytochrome and BAP.

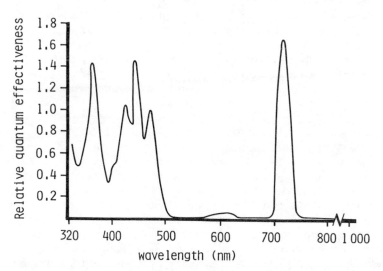

Figure 3.7 Action spectrum for inhibition of hypocotyl elongation in *Lactuca sativa* (from Hartmann, 1967).

The work of Beggs *et al.*[14] illustrates the way in which the HIR disappears during greening. These workers have recorded action spectra for the inhibition of hypocotyl elongation in *Sinapis alba* by continuous illumination. The plants used were either etiolated, WL-grown or were dark-grown plants which had received a particular light treatment prior to continuous illumination. The action spectra showed that B, R and FR (716 nm) were the most inhibitory in etiolated plants but this effect is markedly reduced in plants which had received a pulse of R, which would reduce the total phytochrome content, and in WL-grown plants, which would contain very low phytochrome levels (see below). The action spectra of the greening plants show that B and FR had very little inhibitory effect at all and the peak of inhibition in the R had shifted from 655 to 640 nm, which was thought to be due to chlorophyll screening.

Phytochrome in green plants

Phytochrome is synthesised in the Pr form of the molecule.[15] The Pr form of the molecule is much more stable than the Pfr form. Quail *et al.*[15] show that Pfr is degraded 100 times faster than Pr. However, it was pointed out by Fukshansky and Schafer[16] that the intermediate forms of phytochrome were not subjected to the degradation experienced by Pfr and consequently they may play an important role in preserving phytochrome levels under high fluence rates found in the natural environment. This idea has been exploited by Smith *et al.* (1988),[2] who have examined the degradation rates of phytochrome in four species of etiolated seedlings under high fluence rates of WL. They find that the kinetics of phytochrome degradation indicate two populations of phytochrome, one which is rapidly degraded and the other which is degraded more slowly. These workers suggest that a plausible explanation of these findings is that rapid degradation would be caused by enzymes which were Pfr-specific. The slower degradation would represent degradation of a non-specific nature to which all proteins in the cell are subjected but here is measured by the disappearance of phytochrome forms other than Pfr. These workers also find that over a fluence range of 3 to 1 000 μmol m^{-2} s^{-1} the slowly degraded population in particular is protected at high fluence rates. Smith *et al.* suggest that the existence of much of the total phytochrome as intermediates under these high cycling rate conditions accounted for photoprotection from Pfr-specific degradation. Smith *et al.*[2] also suggest that under the high fluence rates of the natural environment the effective concentration of Pfr would be low and would need to couple with the transduction chain more rapidly than the degradative pathway. Alternatively, phytochrome action would depend upon a mechanism not solely dependent on Pfr.

For many years it was very difficult to investigate phytochrome in green plants for two reasons. Firstly because the spectrophotometric assay for phytochrome could not be used in the presence of chlorophyll which absorbs and fluoresces in the same region of the spectrum as that used to assay phytochrome and secondly because the older techniques for purifying phytochrome were not sensitive enough to deal with the lower levels found in green plants. As techniques have improved, particularly as a result of the development of the immunochemical method for detecting and analysing phytochrome[17] and also the technique for the early elimination of chlorophyll from crude extracts during phytochrome purification,[18] it has become feasible to perform green tissue investigations. It has now been found that phytochrome extracted from green plants (GPP) has a lower molecular weight than phytochrome extracted from etiolated plants (EPP). Tokuhisa *et al.* (1985) show that GPP from *Avena* has a molecular weight of 118 kDa whereas EPP is 124 kDa.[18] It is of course possible that these workers have unearthed another problem with proteolytic enzymes unique to green plants. None the less, only 30 per cent of GPP from oats is recognised immunologically by antibodies raised against EPP. This would suggest that both species are present in green plants and gives rise to interesting speculation as to whether this is related to phytochrome function, whether GPP is produced from a separate gene from EPP, whether GPP arises from EPP and so on. These are questions being actively investigated.

Phytochrome and broad-band irradiation

Although a great deal has been learned from experimentation with etiolated plants using actinic radiation, in the natural environment plants have developed chlorophyll and are subject to radiation throughout the spectrum. Setting aside the problem of phytochrome screening by chlorophyll, which will be dealt with later (p. 54), the behaviour of phytochrome under natural light regimes can be predicted from the absorption spectrum which is shown in Figure 3.4. It is obvious from this figure that both forms of phytochrome absorb significant levels of radiation over a wide range of wavelengths. Two main points should be noted. First, both forms absorb B wavelengths and this leads to problems in identifying the photoreceptor responsible for the mediation of a particular response. If in broad-band irradiation the B fluence and or wavelengths are varied, it is difficult to ascertain whether the effect is exerted via BAP or Pfr/Ptot (ϕ). It is not always appreciated that the effect of Pfr can be enhanced or reduced by B according to the ϕ at which the response is induced. At low fluence rates, it is possible that the supply of photosynthate could limit the magnitude of a phytochrome response. Under these conditions

Table 3.1 Phytochrome photoequilibria under various light conditions.

Sunlight	0.4
Under leaf canopy	0.04–0.2
Incandescent bulbs	0.44
Daylight fluorescent	0.78
Broad band blue	0.22
Red	0.8

(After Bewley and Black 1982). The above data was derived from spectrophotometric measurement of etiolated hypocotyls of either *Phaseolus vulgaris* or *Amaranthus caudatus* by Holmes and Smith (1977) and Kendrick and Frankland (1969) respectively.

the addition of B could indirectly enhance responses. The second point to be noted is that the absorption spectra of Pr and Pfr overlap. In consequence, if 600 nm were supplied, the photons would be absorbed predominantly by Pr; if 700 nm, predominantly by Pfr; and if 690 nm, almost equally by both forms of phytochrome. The phytochrome system must be viewed as a dynamic equilibrium. Actinic R will not entirely convert Pr to Pfr but will produce a photostationary state where $\phi = 0.8$ (i.e. 80 per cent Pfr). In the natural environment the energy absorbed by phytochrome is integrated and the gross effect of various natural and artificial sources on phytochrome in etiolated tissue is shown in Table 3.1.

Phytochrome localisation

Distribution within the plant

The *in vivo* spectrophotometric measurement of phytochrome in etiolated material has been possible for many years[19] and several authors have used this assay to detect the regions of the plant which contain high levels of phytochrome. Much less is known of the distribution of phytochrome in green plants since the spectrophotometric technique cannot be used in the presence of chlorophyll and detection depends upon the use of sophisticated and recently developed techniques such as the immunological detection of phytochrome[17] or microbeam irradiation and biological response.[20] The technique of Jabben and Deitzer[21] of producing achlorophyllous plants by growing them in the presence of a bleaching herbicide has also been useful in this respect. Adding to these difficulties is the fact that the concentration of phytochrome in light-grown plants is an order of magnitude less than that found in etiolated tissue.[22] Even less is known of the localisation of BAP as a result of the lack of a specific assay.

Studies in etiolated material include that of Briggs and Siegelman,[23] who showed that in 5-day-old pea seedlings the vast majority of the phytochrome was associated with the apical bud and the first node, i.e. in the top 2 cm of a plant 12 cm tall. The root, cotyledon and cotyledonary node all had detectable levels of phytochrome.

Tepfer and Bonnet[20] used microbeams of R to irradiate small groups of cells in the roots of *Convolvulus arvensis*. They not only showed that phytochrome was involved in the gravitropic response (see p. 76), but since the response could only be demonstrated by irradiation of cells of the terminal 1 mm of the root tip, they also demonstrated phytochrome location in this area.

The distribution of phytochrome in etiolated oat seedlings was demonstrated first by the spectrophotometric technique[23] and by early immunological techniques[24] and finally by a modified spectrophotometric technique.[25] There is reasonable agreement between these workers in that phytochrome is found in greatest concentration in the tip of the coleoptile. This level decreases down the coleoptile to the node, where it is still quite high, and then falls away rapidly. Low levels of phytochrome are found in the new leaves and below in the mesocotyl. Pratt and Coleman[24] were also able to demonstrate a localisation of phytochrome in the tip of the root. More recent work with etiolated oats[26] indicates that the sites of light perception and the location of the phytochrome which controls mesocotyl and coleoptile growth are separate. Light is piped down the etiolated seedling (see also p. 68).

As was mentioned above, much less is known of the localisation of phytochrome in the green plant. Since the role of phytochrome differs in the mature, green plant compared to that in the etiolated juvenile plant, it is possible that high levels of phytochrome are found in different localities. As was mentioned earlier (p. 30), at the molecular level phytochrome extracted from green plants appears to be significantly different from phytochrome extracted from etiolated plants. EPP has molecular weights between 120 and 127 kDa depending on plant source and this can be shown to be immunologically as well as physically different from GPP. The latter form has a molecular weight of 118 kDa when extracted from oat.[27-29] EPP, otherwise known as Form I, has been detected at low levels in green plants.[28] Tokuhisa and Quail (1987)[30] show that over the first 72 h of etiolated oat seedling development the EPP form increases 200-fold whereas in green tissue the level of this form of phytochrome is relatively constant. The GPP on the other hand increases only twofold over the same period in both light- and dark-grown tissue. Although phytochrome of green plants is present at lower concentration, the Pfr in this situation is more stable. Using the bleaching herbicide Norflurazon, Jabben and Deitzer[21] were able to demonstrate the occurrence of phytochrome in the cotyledons of various dicotyledons, in the primary leaves of oat seedlings and the secondary leaves of *Zea mays*.[30,31]

Localisation of phytochrome within the cell

Many workers, over many years, have addressed the problem of intracellular localisation of phytochrome. These efforts have led to only limited success. Of the early reports of phytochrome association with membranes only that of Haupt (1966)[32] has stood the scrutiny of time and improved techniques. Haupt was able to demonstrate that the phytochrome which was responsible for controlling the orientation of the chloroplast in the green filamentous alga *Mougeotia* was located on or near a membrane close to the limiting membrane of the cell. Although this finding has led to much information[33] it is of limited use in understanding the organisation of phytochrome in the cell of the higher plant.

There is little unambiguous evidence in the literature which shows phytochrome association with membranes. Early reports that phytochrome was associated with various subcellular structures and particles were finally shown to be due to the high concentration of Mg^{2+} used in the extraction procedures. Although light-induced phytochrome pelletability is a genuine phenomenon it is not safe to assume that this reflects associations *in vivo*.

Two areas of research have been moderately successful, if rather contradictory. The first relates to phytochrome association with the organelles. The work of Evans and Smith[34] and Hilton and Smith[35] showed that isolated etioplasts would release GA on a R/FR reversible basis. Nothing more has been reported in this regard since, nor has this effect been reported for chloroplasts. Georgevich *et al.*[36] showed that radio-iodinated phytochrome would bind to mitochondria; this, together with the work of Cedel and Roux,[37] who showed that phytochrome mediated various aspects of the metabolic activity of the mitochondria including NADH dehydrogenase after organelle isolation, made a strong case for phytochrome association with this organelle. The association between phytochrome and mitochondria occurs *in vivo* only after R irradiation. This binding was found to be resistant to proteolysis by trypsin and chymotrypsin, which suggests that the phytochrome is not exposed on the outer membrane of the mitochondria.[38] Indirect evidence also associates phytochrome with the nucleus. Ernst and Oesterheldt[39] show that if isolated nuclei from rye seedlings are added to 124 kDa phytochrome also purified from etiolated rye, then when phytochrome is in the Pfr form the overall transcription rate is increased by 70 per cent. Pr, or partially digested Pfr (114 to 118 kDa), has only weak stimulatory effects on transcription. Mosinger *et al.*[40] were able to obtain similar results with a heterogeneous system of rye phytochrome and oat nuclei.

The second line of investigation has used immunological techniques to demonstrate the location of phytochrome within the cell. The work of Saunders *et al.*[41] demonstrated the distribution of phytochrome in a cross section of etiolated pea epicotyl. They showed that phytochrome was absent or in very low concentration in the epidermal cells with the exception of stomatal guard cells, which contain high levels. The role of phytochrome in the operation of the guard cell may not yet be understood. Present knowledge would indicate that phytochrome is of subsidiary importance compared to BAP in the detection of light by the guard cell (see p. 99). In this work phytochrome-associated fluorescence was detected throughout the cytosol of the cortical cells.

Using dark-grown oat coleoptiles, McCurdy and Pratt[42] show that within 1 to 2 seconds of the beginning of irradiation with R, phytochrome in the Pfr form becomes associated with some sort of binding site and then this sequestered phytochrome becomes aggregated. The sequestered state of phytochrome depends on the maintenance of phytochrome in the Pfr condition. These authors point out the similarity between this process and R-induced pelletability of phytochrome and suggest that the two techniques measure the same event. Attempts to identify the regions where phytochrome aggregations occur have been unsuccessful but it may be said that Pfr does not appreciably associate with readily identifiable organelles such as mitochondria, plastids or nuclei.[22]

Blue-light-absorbing pigment (BAP)

The involvement of BAP in the development of higher plants has been known for many years, perhaps the earliest report being that of phototropic bending by Sachs (1864).[43] However, since B is absorbed by pigments such as phytochrome, which can directly affect development, and chlorophyll, which may have an indirect effect,[44] the existence of a B photoreceptor controlling many aspects of development has not been universally accepted until recent years.

In this area of photobiology there is some young nomenclature with rather vague definitions. The term cryptochrome is used to describe action spectra with characteristic maxima and shoulders in the B region of the spectrum (see below). This term, being derived from cryptogam, indicates the common occurrence of this pigment within this group and some authors use it exclusively for pigments of lower plants with this action spectrum. The term B/UV-A photoreceptor is used to describe B-absorbing pigments in both higher and lower plants. The term indicates the wavelengths which stimulate this pigment and it is particularly useful when B/UV-A is being discussed with reference to the other UV-absorbing pigment, UV-B. Some authors use the term BAP in the broad sense to include a number of molecules which absorb B but are not thought to be involved in controlling developmental processes. The greatest usage of BAP is to describe the pigment or pigments which mediate developmental processes stimulated by B in which the action is not mediated via phytochrome. Ignoring the broader use of BAP, these terms may be largely but not entirely synonymous. Complete information is not available on all known responses to B. Here the term BAP will be used to cover the responses of higher plants stimulated by B which are not thought to be due to phytochrome or any other identifiable pigment irrespective of whether the characteristics of the action spectrum of the response are known.

In higher plants there are at least two developmental pigments which absorb in the UV region of the spectrum. Little is known of the first, known as the UV-B photoreceptor. Its existence was first indicated by Wellman,[45,46] who found that cell suspension cultures of parsley (*Petroelinium hortense*) would only form flavone glycosides when UV-B was part of the radiation. An action maxima was found between 280 and 320 nm with practically no response in the visible region of the spectrum. The involvement of this pigment has since been indicated in a number of systems including anthocyanin synthesis in *Sorghum bicolor*.[47] It may be that this pigment is ubiquitous in occurrence but the difficulties of working with these wavelengths do not encourage research in this area.

By contrast, the B/UV-A photoreceptor is known to be widely distributed throughout the plant kingdom. As the name suggests, BAP absorbs not only in the UV-A region of the spectrum (320 to 400 nm), generally in the region of 370 nm, but also in the B, where it has an action maxima at about 450 nm and shoulders near 425 and 480 nm. Because of such action spectra, some workers would classify this response as one mediated by the hypothetical pigment cryptochrome. Many, although not all, B/UV-A responses have such an action spectrum. The variability of the action spectra in the B region is often said to be due to the environment of the chromophore rather than to an indication of the plurality of the pigment, although it is not yet possible to eliminate the latter possibility. The chromophore environment includes not only parameters such as local pH and ion concentration but also the attachment of other molecules such as proteins.

Although the B/UV-A-absorbing pigment remains unidentified, the flavo-proteins are currently thought to most closely resemble the physiological and biochemical characteristics of this elusive pigment or pigments. The majority of evidence indicting flavines and flavoproteins comes from work with fungi, as does the evidence for carotenoids, which are their main rival.[48]

Senger and Schmidt[49] have reviewed the evidence of the identity and mode of action of BAP and suggest that most of the known data fits the general hypothesis that, as a result of B activation, a membrane-bound flavoprotein undergoes redox reactions which lead to the transmission of the signal down a transduction pathway to the response. Amphiphilic flavines have been synthesised and anchored to artificial phospholipid membranes in model systems by means of long hydrocarbon chains.[50] Such membrane-bound flavins can mediate electron and proton transport resulting in the development of redox and pH gradients.

Demonstrating BAP control

BAP is not the only pigment-absorbing energy in the B region of the spectrum. Chlorophyll has a peak of absorption in the B region of the spectrum (Chl-a near 450 nm, Chl-b near 425 nm) and may under low fluence supply suboptimal levels of photosynthate and limit the ability of the plant to respond to the light environment. Phytochrome has absorption maxima in the B region at 378 nm (Pr) and 392 nm (Pfr). Thus when plants are shown to respond to B it is of prime importance to ascertain which of the above-mentioned B-absorbing pigments is involved.

Mohr (1984)[51] reports that the literature contains three methods which can be used to distinguish between the action of BAP and phytochrome during B irradiations. First, Meijer[52] used a linear variable displacement transducer (LVDT) to measure hypocotyl elongation in dark-grown *Cucumis sativus* var. Venlo Picklin. The LVDT (see p. 69) was used to measure growth in darkness and under B, R and FR irradiations. Meijer found that B inhibited elongation immediately whereas R and continuous FR inhibited elongation only after a lag phase of 30 to 45 min. This work has been amplified by Black and Shuttleworth,[53] who, working with another variety of *Cucumis sativus*, showed that the B response was perceived via the hypocotyl while R was detected largely via the cotyledons. Gaba and Black,[54] using a LVDT, showed that with light-grown plants of this variety, Ridge Greenline, B acted more immediately than R in inhibiting hypocotyl extension (see p. 74).

The second method is that described by Thomas and Dickenson.[55] These workers used low-pressure sodium lamps (SOX) to irradiate their plants. These lamps emit light only in a narrow band around 589 nm (orange-red) and have no B wavelengths. Under these conditions of high photosynthetically active radiation (PAR) and high ϕ it was possible to measure the effects of added B. The fluence rate of the added B was such that its addition to the PAR level was not significant. However, the B was found to cause inhibition of hypocotyl elongation in addition to that ascribed to phytochrome in de-etiolated seedlings of lettuce, cucumber and tomato.

The third method is the light equivalence test of Schafer *et al.*[56] Their report shows that the effect of a light source on phytochrome could be characterised by measuring ϕ and V (the rate of conversion of Pr to Pfr and vice versa). Both these parameters can be measured directly in etiolated tissue. It was reasoned that no matter what the light source used to induce a response, the magnitude of the response would be the same if ϕ and V were the same and phytochrome were the only photoreceptor for the response. Again, the involvement of BAP was demonstrated in the control of hypocotyl extension in a number of species.

Interaction with phytochrome

In the natural environment the three known developmental pigments (phyto-chrome, BAP and UV-B) would all be stimulated. Although little is known of the role of the UV-B, many examples are known where phytochrome and BAP can be shown to influence the same response. Smith has suggested that in shade-intolerant species, at least, BAP measures light quantity and that phytochrome measures light quality.[57] There may be much to be said for this idea but several reports of interactions between BAP and phytochrome indicate that a more sophisticated explanation is required. Mohr (1984, 1986)[51, 58] points out that there are two hypothetical possibilities if two photoreceptors were to interact. First, where both photoreceptors can separ-ately elicit a significant response, then they may act independently upon the response and would not necessarily bring about an additive response. Second, the photoreceptors may act on one another in some way before bringing about an integrated response. In their experiments on the glyceraldehyde 3-phosphate dehydrogenase level in the plastid of milo (*Sorghum vulgare*) seedlings, Oelmuller and Mohr[59] interpret their results as indicating that BAP determines the responsiveness of the system to Pfr. Gaba *et al.*[60] and Attridge *et al.*[61] show that a threshold in terms of B energy must be exceeded to inhibit hypocotyl elongation in de-etiolated *Cucumis sativus*. In the presence of R, the level of the B-response threshold is reduced by at least 30-fold. It is suggested that in this de-etiolated system, at least at low fluence, BAP requires the interaction of Pfr to function.

BAP and Phototropism

Phototropic growth is largely associated with the juvenile organs of plants and is a phenomenon which has attracted an overt amount of attention from plant physiologists. As Firn[62] points out, very few mature plants show any evidence of an earlier phototropic bending and it may be that phototropic growth is an option available to the plant which is only used by the seedlings under certain environmental conditions. Firn suggests that it has received so much attention from the early 19th century to the present day because the dramatic differential growth changes caused are both conceptually and experimentally attractive.

A phototropic response is usually defined as the developing curvature of a growing organ in the direction of, or away from, illumination. Four distinct phases of response can be recognised. First, the perception phase when the photoreceptor is being activated by light. Second, a lag phase during which some kind of physiological asymmetry is established across the organ. The third phase is one of differential growth, often referred to as the curvature phase, and this is followed by the fourth phase, the autotrophic phase, where the organ begins to straighten again.

Perception

Action spectra have always been a popular starting point for understanding a photostimulated process and phototropism is no exception. The action spectra for $10°$ curvature for *Avena* coleoptiles are shown in Figures 3.8a and b. It is

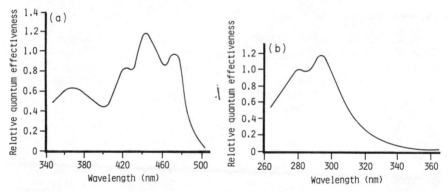

Figure 3.8a Visible action spectrum of 10° phototropism in *Avena* coleoptile (after Thiamann and Curry, 1960). **3.8b** UV action spectrum for 8.6° phototropism in apically shielded *Avena* coleoptiles (after Curry *et al.*, 1956).

not possible to draw comparisons between peak heights in these two figures since in Figure 3.8a the apices were exposed coleoptile[63] and in Figure 3.8b the apices were covered.[64] In both figures the photoreceptor(s) for phototropism has maxima at 290 nm in the UV, and at 475, 450 and 420 nm in the B region. These peaks are not dissimilar to those of the action spectra obtained for phototropic bending in the sporangiophore of the fungus *Phycomyces*.[65] It is largely accepted that phototropism is predominantly a BAP-mediated response. However, it should be noted that there are exceptions to this pattern. The chloronemal filaments of protonemata of the moss *Physcomitricum turbinatum*[66] show practically no effect of B. The maxima found in the R and FR were interpreted as chlorophyll-screening of a Pfr-absorption spectra.

Because the action spectra of phototropism in higher plants shows an action maxima in the B region, early workers in this field considered that R was a safe light which could be used for illumination in pre-experiment manipulations of the plant material. Curry,[67] however, showed that non-directional R pre-irradiation sensitises *Zea* coleoptiles to subsequent irradiations with B. Briggs and Chon[68] went on to show that the pre-irradiation desensitises first-positive response while increasing sensitivity of second-positive response, with the effect that first-negative response becomes occluded by the two positive responses. The effect of pre-irradiation with R can be reversed by FR.

The work of Loser and Schafer[69] and Galland[70] indicates a similar action between R and B in the fungi. These workers found that *Phycomyces* has a small action maxima at 650 nm and that R has a dual role in this system in that when R (605 nm) is supplied with B (450 nm) the phototropic bending is less than when B is supplied alone.

Theories of phototropism

The Cholodny-Went versus the Blaauw hypotheses

This explanation of phototropism comes from the attempts of both Cholodny and Went, who independently tried to find a single hypothesis which explained the control of elongating regions in plants. This theory has received various modifications, qualifications and addenda over the years but is still used as a yardstick when interpreting the phototropic responses for coleoptiles. In short, this theory suggests that light causes the lateral redistribution of auxin at the apex and this redistributed auxin moves down the coleoptile, where the different concentrations of auxins cause differential growth in the region of the bending. It may be pointed out at this stage that in dicotyledonous plants the apex is not the site of photoperception for this response and therefore the hypothesis would need immediate modification for these plants. By contrast, the Blaauw hypothesis is a simpler model and was derived from experiments on sunflower hypocotyls. It is simple in that it suggests that the effect of light is directly upon the cells which respond. That is to say, B light inhibits cell elongation in the responsive area, the response being proportional to the fluence of B received, and causes bending in unilateral illumination as a result of the gradient of B across the shoot. MacLeod *et al.*,[71] working with *Avena* coleoptiles, report results which they are unable to reconcile with the Blaauw hypothesis. By using uneven bilateral illumination they show that the response of the cells depends not only on the fluence rate the cell is experiencing but also on the fluence rate elsewhere in the organ. When $0.03~\mu mol\,m^{-2}\,s^{-1}$ is used as a unilateral illumination it will cause phototropic bending via the inhibition of cells in the elongation zone. However, when the above fluence rate is used on one side and a fluence rate 10 times as high is used on the other, bending takes place towards the higher fluence rate, and the lower fluence has little inhibitory effect on elongation. Furthermore, when a coleptile is irradiated bilaterally and illumination to one side is switched off, then the growth rate on the illuminated side is inhibited even though its light environment would appear to be unchanged. The inescapable conclusion of this work is that these cells are not responding in isolation from one another but have some sort of communication.

Location of the photoreceptor

Meyer[72] produces evidence which suggests that the apex of the coleoptile is primarily responsible for the perception of the light stimulus for phototropism. It was shown that phototropic responses were twice as large when the top $350~\mu m$ were stimulated with a 5-s pulse of B than when the $350~\mu m$ segment below was illuminated. When the top 5 mm of the coleoptile is covered and B irradiation is given for 100 min, small amounts of curvature can be measured.[73]

Fluence-response curves and fluence-response surface

Typical of fluence-response curves are those demonstrated for *Avena* coleoptiles[74] and shown in Figure 3.9. The upper 3 mm of the coleoptiles were exposed to one of three different B (436 nm) fluence rates. The first increase in curvature in this figure is known as the first positive curvature and represents the bending of the coleoptile towards the light. The Bunsen-Roscoe law of reciprocity holds good for the first positive curvature, which is common to all

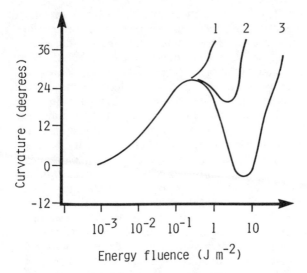

Figure 3.9 Fluence response curves at 436 nm for phototropism in *Avena*. Curves 1, 2 and 3 are 3.84×10^{-4}, 3.84×10^{-3} and 3.84×10^{-2} Wm^{-2} respectively (after Zimmerman and Briggs, 1963).

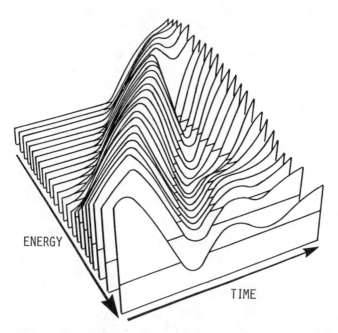

Figure 3.10 Phototropic curvature of *Avena* coleoptiles as a function of B fluence rate (increasing from 10^{-5} Wm^{-2} at back to 12.6 Wm^{-2} at front) and time (increasing from 0.01 s on left to 40 min on right). The coleoptiles had been pre-treated with R and curvature was measured 100 min after end of treatment (drawn from Blaauw and Blaauw-Jensen, 1970).

three fluence rates but does not apply to later responses. The Bunsen-Roscoe law of reciprocity indicates that the response is proportional to the fluence received, i.e. time of exposure × energy. If we consider the effects of the higher fluence rate it becomes obvious that the first positive response gives way to the first negative, which means the coleoptile is beginning to straighten; at higher fluence the coleoptile begins to bend towards the unilateral illumination again and this is known as second positive. As fluence rates decrease (Figure 3.9), then first negative becomes occluded by second positive. To gain as complete an understanding as possible of the effect of light on this system we must consider the three-dimensional model assembled by Blaauw and Blaauw-Jensen[75] (Figure 3.10). From the back left-hand corner of the model fluence rate increases to the front of the model. The time axis runs from the front of the model to the front right-hand corner. If we consider the front of this model at the highest fluence of 12.6 Wm^{-2}, three maxima are obvious, viz. first, second and third positive over 40 min of exposure. As fluence drops, first positive moves along the time axis in accordance with the law of reciprocity. The third positive does not obey this law, does not move along the time axis and gradually merges with second positive and first positive.

Involvement of R with coleoptile phototropic responses

In the interest of producing experimental material which is easy to handle, R pretreatments have often been used and, to a much greater extent, green 'safe lights', which are in actual fact capable of stimulating VLFR. Some inconsistencies in the literature may stem from pretreatments and the use of safe lights. In both experiments mentioned above, red light pretreatments were used. Curry[67,73] showed that the use of R produces a tenfold shift to higher fluences of the first positive and first negative responses and the third positive in the opposite direction. The second positive would seem to disappear if R pretreatment is not supplied. The effect of the R pretreatment varies with the region of the coleoptile to which it is supplied, and Briggs and Chon[68] have shown that there is a correlation between the effectiveness of R in shifting these responses and the distribution of phytochrome. The phytochrome effect is sensitive to very low fluences and may well be classified as VLFR.

IAA and the coleoptile

A library of literature exists which purports to show the Cholodny-Went theory is either right or wrong. It was expected that as radioactively labelled IAA became available it would become possible to distinguish between the destruction of IAA unilaterally and the lateral transport of IAA during unilateral illumination. Furthermore, it was hoped that the polar transport of IAA on the illuminated and unilluminated side could be measured unambiguously. But these hopes have not been realised. Gardner *et al.*[76] show that when exogenous IAA was applied to the apex of the *Avena* and *Zea* coleoptiles a redistribution could be found after first positive stimuli was given, but if the coleoptiles had been pretreated with R then the redistribution did not occur while the curvature still developed. Firn[62] points out that none of the studies demonstrate IAA redistribution during the lag phase or latent phase of the response, which is when it would have to occur to be causal. He also points out that salient measurements of IAA would be those in the region of the elongation zone rather than those diffusing from the two sides of the cut end of the apex. The latent period for *Avena* coleoptile for the second positive response is about 4 min, and large redistributions of IAA should be taking place during this period if the Cholodny-Went hypothesis is correct, but no such redistributions have been reported. The viability of the Cholodny-Went theory depends upon IAA being the limiting factor in the expansion of cells in the elongation zone, and this has not been demonstrated to the satisfaction of all.

Dicotyledons and phototropism

The majority of phototropic research has concerned the monocotyledonous plant and in particular the coleoptile of cereals. Our knowledge of hypocotyl and epicotyl phototropic bending can only be described as fragmentary by comparison. Everett[77] compared the dose response curves for the hypocotyl phototropic bending of radish seedling (*Raphanus sativus*) grown in complete darkness, darkness with a final 24 h period of R and in WL. She reports that whereas first positive can be discerned in the dark- and R-growth conditions, no response could be found to energy in the first positive range in the WL-treated plants, but the plants' response to second positive energies was similar to that to the other treatments.

Some evidence can be found for the Cholodny-Went theory in studies of the hypocotyl of the dicotyledonous seedling. Firn and Digby[78] suggest that phototropic bending occurs when IAA moves out of the cells about to elongate on the irradiated side of the organ. A different view is held by Bruinsma *et al.*,[79] who measured the concentration of IAA in the elongation zone of the hypocotyl of *Helianthus annuus* during phototropic bending. Whereas no redistribution of endogenous IAA was reported, xanthoxin, an abscisic acid-like growth inhibitor derived from the carotenoid violaxanthin, was found in greater concentration on the illuminated rather than the unilluminated side. That this is causally involved with inhibition of cell elongation during phototropic bending is contested by Fern,[62] who reports that phototropic responses are perfectly normal in carotenoid-lacking mutants and in plants treated with inhibitors of carotenoid synthesis.

Directional bending in dicotyledonous seedlings in the natural environment may be a complicated phenomenon. For example, R light can be shown to produce curved growth of the hypocotyl when one of the cotyledons of de-etiolated *Cucumis sativus* or *Helianthus annuus* is masked. Shuttleworth and Black[80] point out that true phototropism may be demonstrated in both these species if the hypocotyl is unilaterally illuminated with B and that the cotyledons play no part the perception of the stimulus. However, it would be interesting to know whether the quasi-phototropic response that can be evoked by masking a cotyledon can be extrapolated to a response to heavy neutral shading or heavy vegetational shade. It could possibly provide the epigeal seedling with a curved growth response in a situation where a cotyledon is shaded but the hypocotyl is experiencing equal multi-lateral fluence. It would seem that in this system and with *Lavatera cretica* a substance which inhibits growth of the hypocotyl is secreted by the cotyledon as a result of illumination.[81] B in de-etiolated *Cucumis sativus* is detected primarily through the hypocotyl rather that the cotyledon.[53] However, with etiolated *Cucumis sativus* at least, Cosgrove[82] has shown a kinetic difference between the B-induced inhibition of elongation growth and phototropic bending. Whereas the inhibition of elongation growth is apparent within a minute or two,[52] the phototropic bending takes several hours to develop. Rich *et al.*,[83] used de-etiolated *Sinapis alba* equilibrated under SOX (see p. 36) irradiated unilaterally with a fluence of B too low to inhibit longitudinal growth but adequate to cause phototropic bending. The tiny seeds of *Scrophularia auricularia* were used to mark the hypocotyl at intervals. Adherence was achieved by the use of a little stop-cock grease. The seedlings were photographed hourly by the light of the SOX lamps and changes in the distances between the *Scrophularia* seeds were measured from enlarged negatives. Their results showed that phototropic curvature was due not only to inhibition of cell growth on the illuminated side, which agrees with the Blaauw hypothesis, but also to stimulation of growth on the shaded side of the hypocotyl, which does not form part of the Blaauw hypothesis.

References for Chapter 3

1   Kendrick, R.E. and Spruit, C.J.P. 1977. Phototransformations of phytochrome. *Photochem. Photobiol.*, **26**, 201–204.

2   Smith, H., Jackson, G.M. and Whitelam, G.C. 1988. Photoprotection of phytochrome. *Planta.*, **175**, 471–477.

3   Rudiger, W. 1986. The chromophore. In: Kendrick, R.E. and Kronenberg, G.H.M. (eds), *Photomorphogenesis in Plants*. Martinus Nijhoff, Dordrecht, 17–32.

4   Viestra, R.D. and Quail, P.H. 1986. The protein. In: Kendrick, R.E. and Kronenberg, G.H.M. (eds), *Photomorphogenesis in Plants*. Martinus Nijhoff, Dordrecht, 35–60.

5   Smith, H. 1983. Is Pfr the active form of phytochrome? *Phil. Trans. R. Soc. Lond.*, **303**, 443–452.

6   Butler, W.L., Lane, H.C. and Siegelman, H.W. 1963. Non-photochemical transformations of phytochrome *in vivo*. *Plant Physiol.*, **38**, 514–519.

7   Smith, H. and Attridge, T.H. 1970. Increased phenylalanine ammonia-lyase activity due to light treatment and its significance for the mode of action of phytochrome. *Phytochem.*, **9**, 477–485.

8   Shanklin, J., Jabben, M. and Viestra, R.D. 1987. Red light-induced formation of ubiquitin-phytochrome conjugates: identified as possible intermediates of phytochrome degradation. *Proc. Nat. Acad. Sci.*, USA, **84**, 359–663.

9   Borthwick, H.A., Hendricks, S.B., Parker, M.W., Toole, E.H. and Toole, V.K. 1952. A reversible photoreaction controlling seed germination. *Proc. Nat. Acad. Sci.*, USA, **38**, 662–666.

10 Raven, C.W. and Spruit, C.J.P. 1973. Induction of rapid chlorophyll accumulation in dark grown seedlings. III. Transport model for phytochrome action. *Acta Bot. Neerl.*, **22**, 135–143.

11 Mandoli, D.F. and Briggs, W.R. 1981. Phytochrome control of two low irradiance responses in etiolated oat seedlings. *Plant Physiol.*, **67**, 733–739.

12 Hartmann, K.M. 1966. A general hypothesis to interpret 'high energy phenomena'

of photomorphogenesis on the basis of phytochrome. *Photchem. Photobiol.* **5**, 349–366.

13　Hartmann, K.M. 1967. Ein Wirkungsspectrum der Photomorphogenese unter Hochenergiebedingungen and seine Interpretation auf der basis des Phytochrome (Hypokotylwachstumshemmung bei *Lactuca sativa*). *Z. Naturforsch.*, **22b**, 1172–1175.

14　Beggs, C.J., Holmes, M.C., Jabben, M. and Schafer, E. 1980. Action spectra for the inhibition of hypocotyl growth by continuous irradiation in light and dark-grown *Sinapis alba* seedlings. *Plant Physiol.*, **66**, 615–618.

15　Quail, P.H., Schafer, E. and Marme, D. 1973. Turnover of phytochrome in pumpkin cotyledons. *Plant Physiol.*, **52**, 128–134.

16　Fukshansky, L. and Schafer, E. 1983. Models of photomorphogenesis In: Shropshire W., Mohr H. (eds.), *Encycl. of Plant Physiol.* NS Vol 16A. Berlin, Springer-Verlag, 69–95.

17　Pratt, L.H. 1984. Phytochrome immunochemistry. In: Smith H., Holmes M.G. (eds.), *Techniques in Photomorphogenesis*. London, Academic Press, 201–226.

18　Tokuhisa, J.G., Daniels, S.M. and Quail, P.H. 1985. Phytochrome in green tissue: Spectral and immunological evidence for two distinct molecular species of phytochrome in light grown *Avena sativa. Planta.*, **164**, 321–322.

19　Butler, W.L., Norris, K.H., Siegelmann, H.W. and Hendricks, S.B. 1959. Detection, assay and preliminary purification of the pigment controlling photoresponsive development of plants. *Proc. Nat. Acad. Sci.*, **45**, 1703–1708.

20　Tepfer, D.A. and Bonnet, H.T. 1973. The role of phytochrome in the geotropic behavior of roots of *Convolvulus arvensis. Planta.*, **106**, 311–324.

21　Jabben, M. and Deitzer, G.F. 1978. A method for measuring phytochrome in plants grown in white light. *Photochem. Photobiol.*, **27**, 799–802.

22　Pratt, L. 1986. Localization within the plant. In: Kendrick R.E., Kronenberg G.H.M. (eds.), *Photomorphogenesis in Plants*. Dordrecht, Martinus Nijhoff, 61–81.

23　Briggs, W.R. and Siegelman, H.W. 1965. Phytochrome distribution in etiolated seedlings. *Plant Physiol.*, 38 Suppl.-5A.

24　Pratt, L.H. and Coleman, R.A. 1971. Immunocytochemical localisation of phytochrome. *Proc. Nat. Acad. Sci. USA*. **68**, 2431–2435.

25　Kondo, N., Inoue, Y. and Shibata, K. 1973. Phytochrome distribution in *Avena* seedlings measured by scanning a single seedling. *Plant Sci. Lett.*, **1**, 165–168.

26　Mandoli, D.F. and Briggs, W.R. 1982b. Photoperceptive sites and the function of tissue light-piping in photomorphogenesis of etiolated oat seedlings. *Plant Cell Environ.*, **5**, 137–45.

27　Tokuhisa, J.G., Daniels, S.M. and Quail, P.H. 1985. Phytochrome in green tissue: Spectral and immunolochemical evidence for two distinct species of phytochrome in light grown *Avena* tissue. *Planta.*, **164**, 321–322.

28　Shimazaki, Y. and Pratt, L.H. 1985. Immunochemical detection with rabbit polyclonal antibody of different pools of phytochrome from etiolated and green *Avena* shoots. *Planta.*, **164**, 333–344.

29　Abe, H., Yamamoto, K.T., Nagatani, A. and Furuya, M. 1985. Characterisation of green tissue specific phytochrome isolated immunolochemically from pea seedlings. *Plant Cell Physiol.*, **26**, 1387–1399.

30　Tokuhisa, J.G. and Quail, P.H. 1987. The level of two distinct species of phytochrome are regulated differently during germination in *Avena sativa. Planta.*, **172**, 371–377.

31　Shimazaki, Y., Cordonnier, M-M. and Pratt, L.H. 1984. Phytochrome quantitation in crude extracts of Avena by enzyme linked immunosorbent assay for phytochrome. *Physiol. Plant*, **159**, 534–544.

32   Haupt, W. 1966. Die Inversion der Schwachlichtbewegung des Mougeotia-chloroplasten: versuche zur kinetik der phytochrom-umwandlung. *Z. Pflanzen-physiol.*, **54**, 151–160.

33   Haupt, W. and Wagner, G. 1984. Chloroplast movement. In: Colombetti G., Lenci F. (eds.), *Membranes and Sensory Transduction*. New York, Plenum Press, 331–375.

34   Evans, A. and Smith, H. 1976. Spectrophotometric evidence for the presence of phytochrome in the envelope membranes of barley etioplasts. *Nature*, **259**, 323–325.

35   Hilton, J.R. and Smith, H. 1980. The presence of phytochrome in purified barley etioplasts and its *in vitro* regulation of biologically active gibberellin levels in etioplasts. *Planta.*, **148**, 428–437.

36   Georgevich, G., Cedel, T.E. and Roux, S.J. 1977. Use of ^{125}I-labelled phyto-chrome to quantitate phytochrome binding to membranes of *Avena sativa*. *Proc. Nat. Acad. Sci. USA*, **74**, 4439–4443.

37   Cedel, T.E. and Roux, S.J. 1980. Modulation of a mitochondrial function by oat phytochrome *in vitro*. *Plant physiol.*, **66**, 704–709.

38   Serlin, B.S. and Roux, S.J. 1986. Light induced import of the chromoprotein, phytochrome, into the mitochondria. *Biochem. Biophys. Acta*, **848**, 372–377.

39   Ernst, D. and Oesterheldt, D. 1984. Purified phytochrome influences *in vitro* transcription in rye tissue. *EMBO J.*, **3**, 3075–3078.

40   Mosinger, E., Batschauer, A. and Schafer, E. 1985. Phytochrome control of *in vitro* transcription of specific genes in isolated nuclei from barley (*Hordeum vulgare*). *Eur. J. Biochem.*, **147**, 137–142.

41   Saunders, M.J., Cordonnier, M-M., Plaevitz, B.A. and Pratt, L.H. 1983. Immunofluorescence visualisation of phytochrome in *Pisum sativum* epicotyls using monoclonal antibodies. *Planta.*, 545–553.

42   McCurdy, D.W. and Pratt, L.H. 1986. Kinetics of intracellular redistribution of phytochrome in *Avena* coleoptiles after its photoconversion to the active, far red absorbing form. *Planta.*, **167**, 330–336.

43   Sachs, J. 1864. Wirkungen des farbigen lichts auf Pflanzen. *Bot. Z.*, **22**, 353–358.

44   Taylor, G. and Davies, W.J. 1988. The influence of photosynthetically active radiation and simulated shadelight on the leaf growth of *Betula* and *Acer*. *New Phytol.*, **108**, 393–398.

45   Wellman, E. 1971. Phytochrome mediated flavone glycoside synthesis in cell suspension cultures of *Petroselinum hortense* after irradiation with ultraviolet light. *Planta.*, **101**, 283–286.

46   Wellman, E. 1974. Regulation der Flavonoidbiosynthese durch ultraviolettesLicht und Phytochrom in Zellkuturen und Keimlingenvon Petersile (*Petroselinum hortense*). *Ber. Dtsch. Bot. Ges.*, **87**, 267–273.

47   Yatsuhashi, H. and Hashimotot, Shimizu, S. 1982. Ultraviolet action spectrum for anthocyanin formation in Broom and *Sorghum* first internodes. *Plant Physiol.*, **70**, 735–741.

48   Horwitz, B.A. and Gressel, J. 1986. Properties and working mechanisms of the photoreceptors. In: Kendrick R.E., Kronenberg G.H.M. (eds.), *Photomor-phogenesis in Plants*. Dordrecht, Martinus Nijhoff, 159–183.

49   Senger, H. and Schmidt, W. 1986. Cryptochrome and UV receptors. In: Kendrick R.E. Kronenberg G.H.M. (eds.), *Photomorphogenesis in Plants*. Dordrecht, Martinus Nijhoff, 137–156.

50   Schmidt, W. 1984a. Blue light physiology. *Bioscience.*, **34**, 698–704.

51   Mohr, H. 1984. Criteria for photoreceptor involvement. In: Smith H., Holmes M.G. (eds.), *Techniques in Photomorphogenesis*. London, Academic Press, 13–42.

52   Meijer, G. 1968. Rapid growth inhibition of gherkin hypocotyls in blue light. *Acta Bot. Neerl.*, **17**, 9–14.

53   Black, M. and Shuttleworth, J.E. 1974. The role of the cotyledons in the photocontrol of hypocotyl extension in *Cucumis sativus*. *Planta.*, **117**, 57–66.

54   Gaba, V. and Black, M. 1979. Two separate photoreceptors control hypocotyl growth in green seedlings. *Nature*, **278**, 51–54.

55   Thomas, B. and Dickenson, H.G. 1979. Evidence for two photoreceptors controlling growth in de-etiolated seedlings. *Planta.*, **146**, 545–550.

56   Schafer, E., Beggs, C.J., Fukshansky, L., Holmes, M.G. and Jabben, M. 1981. A method to check the involvement of additional photoreceptors to phytochrome in photomorphogenesis. *Eur. Symp. Light Med. Plant Dev.*, **9**, 14 Bischofsmais, FRG.

57   Smith, H. 1986. The perception of light quality. In: Kendrick R.E., Kronenberg G.H.M. (eds.), *Photomorphogenesis in Plants*. Dordrecht, Martinus Nijhoff, 196.

58   Mohr, H. 1986. Coaction between pigment systems. In: Kendrick R.E., Kronenberg G.H.M. (eds.), *Photomorphogenesis in Plants*. Dordrecht, Martinus Nijhoff, 547–563. Also see note 62.

59   Oelmuller, R. and Mohr, H. 1984. Responsivity amplification by light in phytochrome-mediated induction of chloroplast glyceraldehyde-3-phosphate dehydrogenase in the shoot of milo (*Sorghum vulgare*). *Plant, Cell and Environ.*, **7**, 29–37.

60   Gaba, V., Black, M. and Attridge, T.H. 1984. Photocontrol of hypocotyl elongation in de-etiolated *Cucumis sativus*. Long term fluence rate-dependent responses to blue light. *Plant Physiol.*, **74**, 897–900.

61   Attridge, T.H., Black, M. and Gaba, V. 1984. Photocontrol of hypocotyl elongation in light-grown *Cucumis sativus*. *Planta.*, **162**, 422–426.

62   Firn, R.D. 1986. Phototropism. In: Kendrick R.E., Kronenberg G.H.M. (eds.), *Photomorphogenesis in Plants*. Dordrecht, Martinus Nijhoff, 367–389.

63   Curry, G.M., Thimann, K.V. and Ray, P.M. 1956. The base curvature of *Avena* to the ultraviolet. *Physiol. Plant*, **9**, 429–440.

64   Thimann, K.V. and Curry, G.M. 1960. Phototropism and phototaxis. In: Florkin M. and Mason H.S. (eds.), *Comparative treatise, Vol I Sources of free energy*. New York, Academic Press, 243–309.

65   Curry, G.M. and Gruen, H.E. 1959. Action spectra for the positive and negative phototropism of *Phycomyces* sporangiophores. *Proc. Nat. Acad. Sci. USA*, **45**, 797–804.

66   Nebel, B.J. 1969. Responses of moss protonemata to red and far-red polarized light: evidence for disc shaped phytochrome receptors. *Planta.*, **87**, 170–179.

67   Curry, G.M. 1957. Studies on the spectral sensitivity of phototropism. Ph.D. Thesis. Harvard University, USA.

68   Briggs, W. and Chon, H.P. 1966. The physiological versus the spectrophotometric status of phytochrome in corn coleoptiles. *Plant Physiol.*, **41**, 1159–1166.

69   Loser, G. and Shafer, E. 1980. Phototropism in *Phycomyces*: a photochromic sensor pigment? In: Senger, H. (ed.), *The Blue Light Syndrome*. Berlin, Springer-Verlag, 244–250.

70   Galland, P. 1983. Action spectra of photogeotropic equilibrium in *Phycomyces* wild type and three behavioral mutants. *Photochem. Photobiol.*, **37**, 221–228.

71   Macleod, K., Digby, J. and Firn, R.D. 1985. Evidence inconsistent with the Blaauw model of phototropism. *J. Exp. Bot.*, **36**, 312–320.

72   Meyer, A.M. 1969a. Versuche zur I. Positiven und zur negativen phototropischen Krummung der *Avena* koleoptile: I Lichtperceptionund Absorptiongradient. *Z. Pflanzenphysiol.*, **60**, 418–433.

73   Curry, G.M. 1969. Phototropism. In: Wilkins M.B. (ed.), *The Physiology of Plant Growth and Development*. New York, McGraw-Hill, 241–273.

74   Zimmerman, B.K. and Briggs, W.R. 1963a. Phototropic dosage-response curves for oat coleoptiles. *Plant Physiol.*, **38**, 248–253.

75   Blaauw, O.H. and Blaauw-Jensen, G. 1970a. The phototropic responses of *Avena* coleoptiles. *Acta Bot. Neerl.*, **19**, 755–763.

76   Gardner, G., Shaw, S. and Wilkins, M.B. 1974. IAA transport during the phototropic of intact *Avena* and *Zea* coleoptiles. *Planta.*, **121**, 237–251.

77   Everett, M. 1974. Dose-response curves for radish seedling phototropism. *Plant Physiol.*, **54**, 222–225.

78   Firn, R.D. and Digby, J. 1980. The establishment of phototropic curvatures in plants. *Ann. Rev. Plant Physiol.*, **31**, 131–148.

79   Bruinsma, J., Fransen, J.M. and Knegt, E. 1980. Phototropism as a phenomenon of inhibition. In: Skoog E. (ed.), *Plant Growth Substances*. Berlin. Springer-Verlag, 444–449.

80   Shuttleworth, J.E. and Black, M. 1977. The role of cotyledons in phototropism of etiolated seedlings. *Planta.*, **135**, 51–55.

81   Schwartz, A. and Koller, D. 1980. Role of the cotyledon in the phototropic response of *Lavatera cretica*. *Plant Physiol.*, **66**, 82–87.

82   Cosgrove, D.J. 1985. Kinetic separazation of phototropism from blue light inhibition of stem growth. *Photochem. Photobiol.*, **42**, 745–751.

83   Rich, T.C.G., Whitlam, G.C. and Smith, H. 1985. Phototropism and axis extension in light grown mustard (*Sinapis alba*) seedlings. *Photochem. Photobiol.*, **42**, 789–792.

4

Seed germination

Introduction

Seeds of many species require the presence of a particular level of Pfr to stimulate germination. This does not mean that all such seeds require light because many contain sufficient Pfr to germinate when imbibed in darkness. This Pfr was formed in the ripening seed and was retained unaltered during the dehydrated existence of the seed. Other seeds have such a low requirement for Pfr that they are stimulated to germinate with the very low level of Pfr provided by FR irradiation, and are capable of development on or near the soil surface, irrespective of the natural light regime. Some seeds require higher levels of Pfr, which can be produced by short irradiation, while others need more complex light treatments to germinate. Such seeds are referred to as light dormant. Although the number of species in the plant kingdom which produce light-dormant seeds are comparatively few, where light dormancy occurs the seeds are provided with a mechanism to detect the suitability of the light environment for germination and development. If a generalisation were to be made it would be that light-sensitive seeds are small and numerous. Light sensitivity is common among ruderals, plants with disturbed habitats.

Although our knowledge of the effect of light on seed germination is now quite detailed it must be emphasised that the control of seed dormancy and germination is a complicated phenomenon. An understanding of light dormancy is only a small part of this phenomenon. Light may also interact with other environmental factors such as temperature or may be a transient form of control which gives way to other mechanisms.

Light within seeds

The light received by the seed may often be filtered through covering structures around the seed. Sometimes these coverings possess chlorophyll and the effect of this on the phytochrome and germination of the seed can be predicted.[1] Apart from action spectra distortion there is little evidence as to the exact nature of light that seeds receive. Small et al.,[2] in their study of action spectra of light induction of germination in lettuce achenes, show that 420 nm has only 1 per cent the effectiveness of 660 nm despite the fact that Pr absorption at 420 nm is 11 per cent more than it is at 660 nm.[3] Although explanations involving phytochrome photoequilibria have been offered,[4] an alternative hypothesis might be that the ineffectiveness of B is due to selective attenuation of B by the achene. Widell and Vogelmann[5] were able to measure the internal light regime of *Lactuca* achenes by inserting fibre optic microprobes, showing that B, 450 nm, was preferentially attenuated by both the pericarp and the seed. This explains at least the less than predicted effectiveness of B acting via the phytochrome system. Whether BAP has a role to play in the photocontrol of germination is not entirely certain. If B attenuation by surrounding structures was shown to be a common effect it might explain any observed lack of involvement of BAP in germination. In most experiments the effects of B on seed germination can be explained in terms of phytochrome stimulation. However, there are cases where this explanation is not entirely acceptable and BAP may be involved. These will be discussed later (p. 53). Natural fluence levels of B are seldom used in germination experiments and it is therefore not yet possible to decide whether BAP is commonly involved as a photoreceptor to stimulate germination.

Perception of light quality and quantity by seeds

The effects of light on seed germination are known as photoblastism. Seeds may be positive or negatively photoblastic, which is to say germination may be stimulated or terminated by light. Today's detailed knowledge of these phenomena tends to make these terms redundant. Light is now known to affect seed germination in the following ways:

1) stimulation of germination by photoperiods of either long or short days
2) stimulation of germination by short periods of light
3) stimulation of germination by long periods of light
4) inhibition of germination by long periods of light.

Photoperiodic responses

The control of germination by photoperiod has been known for many years, as has the role of phytochrome in photoperiodic phenomena. Black and Wareing (1955)[6] showed that the seeds of *Betula pubescens* were stimulated to germinate by long days (LD). Fewer than 8 h of light in each 24-h cycle result in less than 40 per cent germination. This level increases linearly to over 90 per cent as the photoperiod increases to 20 h in each 24-hour cycle. Interaction of the photoperiodic response and temperature is marked. *Chenopodium botrys* requires short days (SD) to germinate at temperatures between 10 and 20°, but between 25 and 35° LD are required. Light of low R : FR reduces germination in both long and short days. End-of-day FR treatment, or at least end-of-day low R : FR, considerably reduces the level of germination in SD but, as might be predicted from a knowledge of flowering and photoperiodism, similar treatment in LD (18 to 20 h) is not effective.[7]

Examples of SD germinators are *Tsuga canadensis*, *Veronica persica* and *Puya berteroniana*.[8] Each shows a critical daylength below which germination is stimulated. As with other photoperiodic phenomena (see Chapter 6), the level of Pfr and the length of the dark period, rather than the length of the light period, has been shown to be of prime importance. It may be that if the seeds, reportedly photoperiodic in their germination requirements, were re-examined in the light of today's understanding of the light requirements of germination, some of them would require re-classification (see below, Long-period inhibition, Long period stimulation).

Short-period stimulation

This response has been known since the 19th century but it was not until the work of Borthwick, Hendricks, Parker, Toole and Toole (1952)[9] that the involvement of phytochrome was revealed. Using the light-sensitive variety of lettuce Grand Rapids, these workers performed what is arguably the most important experiment in the history of phytochrome. They showed that seeds imbibed in darkness will respond to a short exposure of low-energy light. R (660 nm) was shown to be the most efficient in promoting germination, although irradiation with a variety of wavelengths capable of absorption by the Pr form of the phytochrome molecule would serve to promote germination above the level of the dark control. If the seeds were irradiated with FR subsequent to the R irradiation, germination was reduced, close to that of the dark control. It is necessary for the Pfr form of phytochrome to exist for a particular period of time to initiate the response. This time is usually referred to as the escape time and it varies between responses, between species and with temperature (see also Germination and Sunflecks, p58). Sensitivity to short irradiations is now known to be a common phenomenon among light-sensitive seeds, e.g. *Senecio vulgaris*,[10] *Lycopersicon esculentum*[11] and *Sinapis arvensis*.[12]

There always seem to be exceptions to biological rules, and the exception to the rule that Pfr stimulates germination may have been supplied by Hilton[13,14,15] from her studies on *Bromus sterilis*. The seeds of this species appear to be inhibited from germinating by Pfr and at present it is unique among the species investigated. Freshly harvested seed (50 per cent water) required 3 to 4 weeks to attain maximum germination. The range of germination could be increased in these seeds by daily 8-h irradiations with R. Seeds which have been dried (10 per cent water), on the other hand, attained full germination in the dark after 4 days, but daily 8-h irradiations with R inhibit germination. Total modification of the response to high Pfr depends upon drying the seed at alternating temperature. Hilton[15] suggests that these responses may be of ecological importance. In England, seeds of *B. sterilis* reach maturity during July when day temperatures reach 28° and night temperatures drop to 6°. If moist seed is shed, this occurrs early in the season and the seed is liable to fall beneath the canopy of surrounding vegetation. These seeds have a light requirement which encourages germination only below gaps in the canopy. Seed shed later in the season would have experienced alternation of temperature during dehydration of the parent plant. This seed has developed photoinhibition and requires burial before germination can ensue. Burial also ensures the seed a degree of moisture which it might not enjoy on the soil surface at this time of the year. Pretty though this scenario is, the mechanism which provides it is not agreed upon by all workers in this area. Ellis *et al.* (1986)[16] do not agree that the inhibition of germination in *B. sterilis* and *B. mollis* is due to an unusual effect of Pfr. These workers showed that germination was progressively inhibited with increasing fluence and increasing photoperiod and suggested that the inhibition of germination might result from the HIR, which is known to inhibit germination in a number of species (see below). The differences between the findings of these workers and those of Hilton may be due to collection of genetically different batches of seed from different locations, or to the collected seeds having experienced factors in the natural environment which prime them to respond either to short irradiations or to high irradiance. Although the findings of Ellis *et al.*[16] are in keeping with our understanding of the inhibitory effects of light on germination (see below), it cannot be denied that Hilton[13] demonstrates partial reversal of inhibition of germination by low-energy R.

Long-period stimulation

Seeds which require long periods of light to germinate may not at first sight lend themselves to an explanation which includes phytochrome control. However, the long-period irradiations may be replaced by intermittent irradiations in many of the species involved, e.g. *Epilobium cephalostigma,*[17] *Kalanchoe blossfeldiana*[18] and *Pawlonia tomentosa.*[19] Frankland has suggested[20] that in a single batch of seeds some individuals respond to short irradiations while others require long treatments. In a batch of *Plantago major*, it was reported that 35 per cent of the seeds responded to a short R treatment, but the overall germination could be increased to 90 per cent by prolonged irradiation for 48 h with R or by 5-min R pulses on a daily basis. The 35 per cent of the population which responded to short irradiation soon after imbibition indicates that phytochrome has a controlling influence early in the germination process, whereas that percentage stimulated by 5-min R treatments on a 24-h basis suggests that, for part of the population, the presence of Pfr is also required to stimulate a later process.

Long-period inhibition

Kendrick and Frankland[21] reported that *Amaranthus caudatus* germinates readily in the dark but can be inhibited from doing so by either prolonged FR or WL, i.e. by prolonged periods of either low or high Pfr. To promote germination, the seeds held in FR must be treated with R before being placed in darkness, whereas those held in WL need only to be transferred to darkness to promote germination (see Figure 4.1). This effect of red is reversible by a pulse of FR. The inhibitory effect of light depends not only upon length of irradiation but also upon fluence rate. The most effective wavelength for inhibition of germination is 720 nm and may well involve the HIR. These effects have been further investigated with *Sinapis arvensis*.

The seeds of *Sinapis arvensis* have low germination in darkness and can be stimulated to germinate by a short exposure to R. These seeds show maximum responsiveness to R about 6 h after sowing.[12] If these seeds are first exposed to R and then exposed to long irradiations of either FR or WL they do not germinate in subsequent darkness. Even if R is given prior to the dark period the seeds will not germinate (see Figure 4.2). Bartley and Frankland have investigated this situation further[22] and have shown that the inhibitory effect of WL and FR is fluence-rate-dependent. They suggest that there is an antagonism between the promotive effect of Pfr and the inhibitory effect of

Final percentage germination

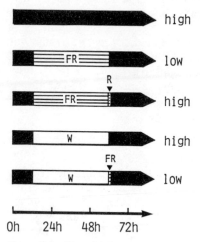

Figure 4.1 Photoinhibition of seed germination in *Amaranthus caudatus* (redrawn from Frankland, 1986).

high fluence rates which can be explained if the HIR operates via the cycling rate of phytochrome, which is fluence-rate-dependent. Bartley and Frankland are able to demonstrate that germination in this species is maximised under conditions where cycling rate is low and ϕ is high.

Cucumis sativus has a seed which germinates well in the dark as a result of Pfr retained in the dried seed. These seeds can be inhibited from germinating by a short period of FR until 40 h after sowing. Beyond this point the germination has escaped from the control of Pfr but can still be inhibited by light by increasing the cycling rate of phytochrome.[23]

Final percentage germination

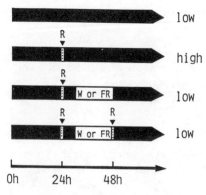

Figure 4.2 Photoinhibition of seed germination in *Sinapis arvensis* (redrawn from Frankland, 1986).

A significant piece of evidence which suggests a role for HIR in seeds germinating in the natural environment is supplied by Gorski and Gorska.[24] These workers subjected the seed of *Lactuca sativa* cv. Vanguard to prolonged irradiations of natural daylight attenuated through various neutral-density and interference filters. The seeds of this variety were regarded as light insensitive because they would germinate well in darkness, in diffuse WL and after short FR. At low fluence rates the R : FR was shown to be the determining factor for germination whereas at high fluence rates of natural daylight the seeds were inhibited from germinating in what was interpreted as a typical HIR, where the FR content of the light was shown to be responsible. Under natural conditions, germination could be promoted in the near-open situation at low fluence and moderate R : FR (50 per cent germination at R : FR of 0.5), but as the fluence increases, the inhibitory effects of HIR begin to dominate. This may mean that although these seeds require high R : FR for high germination rates and may be prevented from receiving such stimulation by shallow burial, they also require a degree of neutral shading to alleviate HIR inhibition. This could be selectable since a degree of neutral shade might also supply necessary moisture. Although no data were presented in this paper, these workers state that similar results have been obtained with seeds of three lettuce varieties, viz. Mays Queen, Grand Rapids and Nochowska, and with *Lactuca serriola* and *Arabis pumila*. Although *Hordeum vulgare*, *Sinapis alba* and *Mesembryanthemum cristallinum* were found to germinate equally well in all spectral bands irrespective of R : FR, small inhibitions of germination were noted at high irradiance of natural daylight.

BAP and germination

The simplest involvement of B in germination occurs via the phytochrome system in seeds which are sensitive to short irradiations of R. If such seeds require a high ϕ to germinate, then B (400 to 500 nm), which generates a ϕ in the region of 0.25, is capable of lowering the germination. On the other hand, if only a low ϕ is required for germination, then B is capable of stimulating germination. Comparatively short irradiations would typify these responses and they can be explained by the phytochrome system.

However, long irradiations of high fluence B also cause responses in seeds and these may occur either via HIR or via BAP. Gwynn and Scheibe[25] present an action spectrum for the inhibition of the germination of lettuce cv. Grand Rapids. These seeds were imbibed for 2 h in the dark and irradiated with R before being returned to darkness for 6 h. By this stage, germination would have escaped control by Pfr. The seeds were then exposed to equal quanta of various wavelengths. Peaks of inhibition appeared between 440 and 500 nm in the B and between 710 and 720 nm in the far red. The B response, like that of FR, requires prolonged exposure and is fluence-rate-dependent. The simplest explanation might be that both light qualities operate through the HIR. However, models of phytochrome action[26,27] which suggest that phytochrome is the sole photoreceptor for the HIR do not explain the efficiency of UV and B wavelengths in action spectra to the satisfaction of all.[4] Further evidence which might indicate the involvement of BAP in germination comes from the work of Malcoste *et al.*[28] These workers published a comparison of the B region action spectrum of the inhibition of germination of *Nemophila insignis* and the effect of these wavelengths on the Pfr/Ptot. The coincidence of these curves is very poor and indicates that B does not act through phytochrome at least for germination in this species. There is not, as yet, sufficient evidence in the literature to suggest that the involvement of BAP in germination is common.

Light effects on the maturing seed

The imposition of dormancy on some light-sensitive seeds can be shown to occur while the seed is still attached to the parent plant and approaching maturity. Dormancy can be classified as photoperiodic or chlorophyll screening.

Photoperiodic effects

Lona (1947)[29] showed that dormancy was initiated by long days (16 to 18 h) in *Chenopodium amaranticolor*. Under this photoperiod small seeds develop with thick seed coats. These seeds are dormant. A photoperiod of 6 to 8 h results in the development of larger thin-coated seeds which are less dormant. In a more detailed study Pourrat and Jaques[30] showed that the photoperiod around the time of flowering was critical in affecting the seed coat thickness and depth of dormancy in *Chenopodium polyspermum*. *Ononis sicula* is induced to produce thin-coated green or brown seeds by short days. These seeds imbibe with ease and germinate readily. In long days larger seeds are produced which have thick yellow seed coats which retard water uptake and delay germination.

Chlorophyll screening

As may be deduced from the absorption spectrum of chlorophyll (Figure 3.2), seeds which dehydrate while surrounded by tissues containing chlorophyll will be irradiated by light of a different quality than that of seeds dehydrating in achlorophyllous surroundings. In particular, the chlorophyll screen will deplete the red wavelengths and provide the seed with a light environment of low R : FR, thus producing low ϕ. Cresswell and Grime[1] suggested that the seeds surrounded by chlorophyll during dehydration are liable to contain too low a ϕ to initiate germination for many species in the appropriate season and will require light stimulation to germinate. On the other hand, seeds which dehydrate in the absence of a chlorophyll screen will dehydrate with a high level of Pfr and will germinate in the appropriate season without a light requirement. These workers have been able to correlate the dark germination level of a number of species with the chlorophyll content of the extra-embryonic tissue.

Similar effects have been shown in *Arabidopsis thaliana*[31] and *Cucumis*.[32] In these species the development of dormancy depends upon the quality of the light regime prevalent during seed ripening on the parent plant. This phenomenon may well be of ecological importance because the seeds which develop under canopy conditions require light to stimulate germination either in a suitable location or in a suitable season.

Light and seed burial

Soil has very opaque qualities (see Chapter 2). All soils examined so far transmit less light than do sandy loams[33] and artificial germination media,[34] but even these examples only transmit enough light to stimulate germination in the top few millimetres. It is not surprising, therefore, to find that small quantities of topsoil contain many thousands of seeds which are dormant as a result of the absence of light. Brenchley and Warington (1930)[35,36] estimated the size of the seed bank in an acre of arable land (approx. 0.5 ha) to be as high as 158×10^6 in southern England, whereas Milton (1936)[37] gives a figure of 74×10^6 for mid-Wales. Wesson and Wareing showed that disturbance of soil which had been under pasture for six years resulted in a flush of germination over the following 4 weeks.[38] Under laboratory conditions 10 per cent of seeds collected from such soil germinated in darkness but when this was repeated in the field, no seedlings emerged. In the field the most common species to emerge after disturbance were *Sinapis arvensis*, *Polygonum aviculare* and *Veronica persica*. After 4 weeks these species occurred at a frequency of 1 780 seedlings m^{-2}, whereas before the disturbance the most common species occurred at 33 seedlings m^{-2}. The seed which germinates under these conditions is not necessarily the seed dispersed in the previous season. The terms primary and secondary dormancy are often used in this area of work, and Karssen[39] defines primary dormancy as that which prevents germination on the parent plant during maturation and immediately after shedding, whereas secondary dormancy develops in seeds after harvest or dispersal. Under field conditions, secondary dormancy has often been observed in buried seeds of many wild species, and the dormancy in these seeds often follows cyclic changes which are associated with the seasons. The seeds of summer annuals, such as *Polygonum persicaria*, develop secondary dormancy during the spring as soil temperatures increase. Dormancy is broken by low winter soil temperatures, and if germination conditions are not correct, this cycle has been shown to repeat itself in the following year.[40] With winter annuals such as *Veronica hederifolia* secondary dormancy is broken by high temperatures and moist atmospheres of the summer months.[41] Cyclic behaviour of germination has been followed over two years in seeds of *Senecio jacobea*, *Chenopodium album* and *Sisymbrium officinale*.[40] *C. album* behaves as a winter annual, whereas the others behave as summer annuals. *Chenopodium bonus-hendricus* does not show cycles of secondary dormancy, but when exhumed, both *Chenopodium* species show large differences in response to light according to the time of year.

Pons[42,43] has investigated the germination of *Cirsium polustre* seeds under ash (*Fraxinus excelsior*) canopies. When *C. polustre* seed is freshly dispersed it is positively photoblastic and is inhibited from germination in this habitat by the low R : FR supplied by the ash canopy. By the time the ash canopy falls, the seed has lost its sensitivity to light, and as winter approaches the conditions for development become unfavourable, increasing the probability that the seed will join the soil seed bank. Seed which is to survive normally becomes buried perhaps by falling into cracks. Seed which remains on the surface is heavily predated. When winter coppicing (felling) occurs, the concomitant soil disturbance may expose the seed to light of a high R : FR. Although winter temperatures are too low to permit germination, the seed regains its light sensitivity and the Pfr produced stimulates germination as soon as temperature permits. If winter felling does not occur, as would be likely in a more natural situation, then the light-sensitive seed remains buried, and even when temperatures are favourable the seed remains dormant.

Photocontrol of seed germination is often regarded as a strategy which ensures that small seeds only germinate at a depth which will allow the seedling to emerge into light of the appropriate quality before the food reserve is exhausted. Studies have been carried out using *Sinapis arvensis* and *Plantago major*, planting seed at different depths in well-aerated fine seed compost.[34, 44] As may be seen in Figures 4.3a and 4.3b, a close correlation between decreasing germination and increasing depth was found. Seeds which germinate at 4 mm in these experiments do so with less than 0.01 per cent of the incident radiation at the soil surface. This would be approximately 0.2 μmol m^{-2} s^{-1} on a summer day with clear sky at temperate latitude. In this situation, phytochrome is acting as a simple light detector and it may be suggested that the sensitivity of phytochrome to low fluence allowed its selection for this response in preference to BAP. From the fluence rate at 6 mm it is possible to calculate that the low percentage of seeds which still germinate are not associated with Pfr stimulation.[34]

More recently, Bliss and Smith[45] have investigated light passing through horticultural sand of various particle sizes. They point out that light passes through voids in the matrix structure and through particles and aggregates and is also reflected by these particles. The overall nature of the transmitted light depends upon the relative proportions of these components, and they suggest that the light received by a particular seed depends very much on its exact location. These workers examined the effect of burial on seven species. Of these, *Plantago major* shows a more or less linear decline in seed germination

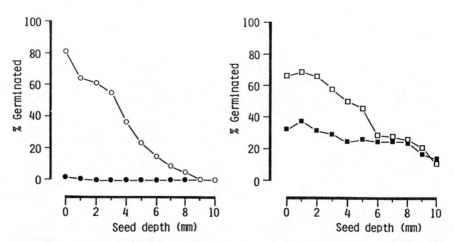

Figure 4.3a Effects of burial on seed germination. Seeds of *Sinapis arvensis* were buried at different depths in soil and exposed to sunlight (open symbols) or left in darkness (closed symbols). **4.3b** A similar experiment to above but using seeds of *Plantago major*. Light exposed seeds (open symbols) seeds left in darkness (closed symbols) (redrawn from Frankland, 1981).

with depth of burial, germination being close to 0 per cent at 10 mm, whereas *Chenopodium album* and *Rumex obtusifolius* lose germination in a threshold manner at 2 mm and 4 mm, respectively. There may be a small promotion of germination with *Chenopodium* when buried very close to the surface. *Cecropia obtusifolia*, a tropical tree, loses germination gradually between 4 and 10 mm. *Galium aparine* and *Amaranthus caudatus* germinate poorly on the surface but when buried reach a maximum level of germination at 2 and 4 mm, respectively. *Digitalis purpurea* is insensitive to depth. Bliss and Smith suggest that the germination at depth of *D. purpurea*, *G. aparine* and *C. obtusifolia* might be explained by an absence of dark (thermal) reversion, allowing these species to develop Pfr levels in their buried seeds over a long period of very low illumination. The increase in germination with depth demonstrated with *G. aparine* and to a lesser extent with *C. album* might be explained by the hypothesis of Bartley and Frankland[22] (see p. 51), who suggest that promotion of germination in seeds of *S. arvensis* is achieved by low cycling rates produced by low fluence rate, coupled with high levels of Pfr. Similar responses have been demonstrated with *Senecio jacobea*[46] and may now be explained in a similar fashion. Thus, *S. arvensis*, *S. jacobea*, *C. album* and *G. aparine* may be inhibited in full sunlight by the cycling rate engendered but this is alleviated by shallow burial. The germination of *A. caudatus* has been previously explained (p. 51) and the photoinhibition is relieved by soil shading.

The above results have recently been reviewed by Tester and Morris (1987),[47] who infer that some of these conclusions could have been strengthened if the total fluence received by the seeds had been recorded. These fluences varied enormously since some seeds were assayed over 3 days and others over 21 days according to the time taken for the controls to reach maximum germination.

Germination in canopy shade

As we have seen previously (p. 16), daylight which passes through a vegetative canopy is selectively attenuated, particularly by chlorophyll. Canopy light is characterised by a lower fluence rate throughout the visible spectrum when compared with the open situation and is particularly depleted in the blue and red wavelengths. Although about 45 per cent of FR is reflected by the leaf surface, a similar amount is transmitted through the leaf, which is why canopy light is relatively rich in FR and has a low R : FR. Exposed to light of this quality, seeds on the surface of the soil or in the top few millimetres of soil will contain low ϕ.

A survey of light-sensitive seeds[48] shows that the majority of positively photoblastic seeds are inhibited from germinating by canopy light, whereas only 60 per cent of negatively photoblastic seeds are affected and only 40 per cent of light-indifferent species. The insensitivity of germination of many species to canopy light may be because these species are able to germinate at Pfr levels lower than that produced by canopy light. Bewley and Black[49] list several species which are shown to germinate at ϕ lower than that found in canopy light.

Plantago major has been grown under canopies of *Sinapis alba* seedlings at various seedling densities.[34,44] It was found that when the light received by the *Plantago* seeds fell below a ϕ of 0.57, germination was reduced. At a ϕ of 0.2 the germination was reduced to that of the dark control. *Plantago* seedlings are very small, and whereas Pons[50] agrees that the ϕ germination requirement is probably a device to prevent the germination of such seeds in the vicinity of established plants, he points out that other environmental factors interact with the R : FR. The inhibitory effect of low R : FR was reduced if the plants were held at high temperature, or if nitrate were made available. Reduced water potential tended to increase the inhibitory effect of low R : FR. Pons argues that the germination of newly shed *Plantago* seed in late summer is to some extent reduced by the high temperature requirement and by the low R : FR if they land under canopy.

Those seeds, which rehydrate containing high levels of Pfr retained from the ripening process on the parent plant (see above), may become buried, and if buried to a sufficient depth (p. 56) the high Pfr level will not be reversed by the canopy light but will initiate germination. Few species exploit this ecological niche in temperate deciduous woodland but undergrowing species are more frequent beneath grass canopies.

Various degrees of vegetational shade exist which will reduce ϕ from the 0.54 value of the open situation to 0.15 under dense vegetational shade.[51,52] These ϕ values are measured in etiolated tissue or calculated by modifications of Hartmann's formula.[26] It is of course impossible to measure ϕ in green material. As was mentioned earlier, Bewley and Black[63] have drawn together a list of species and their germination requirements in respect of ϕ. When more species have been examined a correlation may emerge between the lowest tolerated ϕ value for germination with the seedlings' ability to survive in the habitat that imposes these conditions. The work of Karssen[53] shows that *Chenopodium album* will tolerate a low ϕ of 0.16 to germinate, whereas lettuce requires a ϕ of 0.4. Of these two species, *Chenopodium* would be thought to be more capable of developing in a shady habitat of limited canopy height. Other examples could be cited to agree with this rule but whether a generalisation may be made is a matter which awaits further research. Both these species display a certain amount of tolerance in their light requirements for germination on a population basis. *Lactuca* germinates to 50 per cent under a ϕ of 0.4, but the other 50 per cent requires a ϕ of 0.6. Similarly, *Chenopodium* produces 50 per cent germination at 0.16 but requires 0.3 to promote the remainder. It may well be a successful strategy among plants producing a large number of small seed to produce them with a variety of ϕ requirements to enable them to exploit a range of habitats.

Germination and sunflecks

As was seen in Chapter 2, sunflecks are not homogeneous areas of daylight penetrating the canopy but have regions of low and high R : FR. These non-homogeneous patches of light are characteristically transient in nature. They may last from a few seconds to a few hours.

The first aspect of seed response to sunflecks to consider is escape time. All LFR of phytochrome have an escape time, which is the time that Pfr must exist to initiate a response. Pigment may be reversed by FR after this time, but the response cannot be prevented. Obviously seeds provided with escape times of several hours have a device which allows them to ignore transient light signals that would encourage germination in an otherwise unsuitable light environment.

Escape time for germination varies between species, with the period of imbibition and with temperature. Duke *et al.*[54] show that if seeds of *Portulaca oleracea* were imbibed in the dark for varying periods between 24 and 72 h, then escape times increased with increasing periods of imbibition. In the natural environment, a seed which falls beneath canopy or into a crack in the soil surface would experience imbibition substantially earlier than photostimulation, and this effect would be saturated and escape times would be longer rather than shorter. *Lactuca sativa* shows a 50 per cent escape time of 10 h at 20° but of only 3 h at 25°.[55,56] For some species such a scenario might be disadvantageous. In equatorial climates sunflecks of short duration would not stimulate germination of two species of pioneer trees of tropical rain forests. Vazquez-Yanes and Smith[57] show that *Cecropia obtusifolia* and *Piper auritum* require long periods of R repeated on a daily basis to promote germination (see Long-period stimulation, p. 51).

Parts of the sunfleck with a high R : FR will obviously stimulate Pfr production in mostly long-term imbibed seeds on or near the soil surface. When the sunfleck disappears the seed will receive the R : FR of the canopy environment. If the seed requires a high ϕ to germinate, the Pfr content may fall below the threshold of germination. Whether or not such a seed would germinate would then depend upon the escape time compared to the period of the sunfleck.

Light and temperature interactions

Many seeds which respond to short-period irradiation do so over a wide range of temperature, but in others light and temperature interact in a number of ways which affect seed germination. Most of these interactions are poorly understood.

One interaction which is fairly clear is the effect of temperature on the existence of Pfr. Little work has been done on the direct measurement of Pfr reversion in seeds because of their opacity but measurements have been obtained in two seeds, albeit gymnosperms.[58] In these seeds, *Pinus nigra* and *P. palustris*, much of the Pfr formed by 5 min of R (about 80 per cent) reverts in the first hour of darkness at 24° and reaches a 20-per-cent (0.2 ϕ) level between 6 and 12 h. This process could be considerably important during the dehydration which accompanies seed ripening. However, seeds are more likely to dehydrate during the day while temperatures are high, passing a critical water content which effectively captures Pfr for the germinsation process. These seeds are more likely to retain a high ϕ than seeds which dehydrate during the hours of darkness, although low night temperatures could retard the loss of Pfr. The truth of this conventional wisdom depends upon either the ϕ developed by moonlight and starlight being inactive so far as the phytochrome system is concerned, or the developed ϕ being below the threshold for germination in a particular species (see Chapter 2). Indirect evidence on the reversion of Pfr in seeds is available from physiological experiments by several workers. Taylorson and Hendricks,[59] for example, show that the half life of Pfr increases with decreasing temperatures.

Since reversion is faster at high rather than low temperatures, it cannot be used to explain the variation in photostimulated germination in *Lactuca sativa* cv. Grand Rapids,[49] where dark germination is high at low temperatures and the added effect of R can only be small, whereas at high temperature (25 to 30°), dark germination is low and the stimulation by R is marked. The seeds of *Betula pubescens* produce high levels of germination over a wide range of temperatures under long days.[6] If these seeds are kept in darkness, then a short R treatment is effective at high temperature (20 to 30°) but R is ineffective in promoting germination between 10 and 20°. The germination of *Barbarea* is greatly stimulated by short irradiations at low temperature, beginning at 16°. At 25° the dark control is stimulated to germinate and is as high as the light treatment at 30°. Explanation of these phenomena may become apparent when the initial action of Pfr in these seeds is understood.

Temperature in the natural environment undergoes diurnal variation, and it was reported by von Liebenberg (1884)[60] that the seeds of *Poa annua* were stimulated to germinate by alternation of temperature. Steinbauer and Grigsby (1957)[61] tested 85 species, mainly arable weeds and crop species, and found that over 70 per cent of these seeds could be stimulated in their germination by subjecting them to alternating temperature treatment. Toole *et al.*[62] reported an interaction between light and alternating temperature in the stimulation of germination of *Nicotiana tabacum*. Bewley and Black[63] provide a list of species which have been investigated by various workers from the point of view of germination under alternating temperature. In general, these results show that alternating temperature can largely or partially replace the light requirement necessary to break dormancy in seeds. It has been reported that alternating temperature and light stimulate maximum germination,[64] but more recent publications[65] show that other factors are often required for maximum response. *Rumex obtusifolius* and *R. crispus* require 4 to 10 cycles of alternating temperature in the light where the difference is 5° or more and the lower temperature is above 15° and the higher temperature is below 25°. The magnitude of the response was shown to vary with site and year of collection and to increase slowly with dry storage. This variability has been further studied by Probert *et al.*[66] who report that the proportion of the population of *Dactylis glomerata*, which is sensitive to alternating temperature, varied from year to year and, within a single batch, varied with period of storage. Probert *et al.*[67] show that *Ranunculus scleratus* seeds are characterised by a stringent requirement for alternating temperature, and they observe that this appears to be a characteristic of wetland species. It is of course possible that alternation of temperature provides the seeds of such species with a device to detect the suitability of their rather aqueous environment for germination. Water being a poor conductor of heat, suitable temperature fluctuations would not occur in bodies of water of a critical size. Maximum germination resulted with *R. scleratus* when intermittent pulses of R were given together with daily 4 h temperature shifts from 16 to 26° and when either nitrate or thiourea was present. A similar result was reported by Goudey *et al.*[68] who show that high levels of germination of *Sinapis arvensis* can be achieved by combining the effects of changing temperature and irradiation in the presence of nitrate and ammonium chloride.

References for Chapter 4

1   Cresswell, E.G. and Grime, J.P. 1981. Induction of a light requirement during seed development and its ecological consequences. *Nature,* **291**, 285–287.

2   Small, J.G.C., Spruit, C.J.P., Blaauw-Jensen, G. and Blaauw, O.H. 1979a. Action spectra for light-induced germination in dormant lettuce seeds. I. Red region. *Planta.*, **144**, 125–131.

3   Mancinelli, A. 1986. Comparison of spectral properties of phytochromes from different preparations. *Plant Physiol.*, **82**, 956–961.

4   Cone, J.W. and Kendrick, R.E. 1985. Fluence response curves and action spectra for promotion and inhibition of seed germination in wild type and long-hypocotyl mutants of *Arabidopsis thaliana. Planta.*, **163**, 44–54.

5   Widell, K.O. and Vogelmann, T.C. 1988. Fibre optic studies of light gradients and spectral regime within *Lactuca sativa* achenes. *Physiol. Plant.*, **72**, 706–712.

6   Black, M. and Wareing, P.F. 1955. Growth studies in woody species. VII. Photoperiodic control of germination in *Betula pubesens. Physiol. Plant.*, **8**, 300–316.

7   Cumming, B.G. 1963. The dependence of germination on photoperiod, light quality, and temperature in *Chenopodium spp. Can. J. Bot.*, **41**, 1211–1213.

8   Toole, V.K. 1975. Effects of light, temperature and their interactions on the germination of seeds. *Seed Sci. Technol.*, **1**, 339–396.

9   Borthwick, H.A., Hendricks, S.B., Parker, M.W., Toole, E.H. and Toole, V.K. 1952. A reversible photoreaction controlling seed germination. *Proc. Nat. Acad. Sci.*, **38**, 662–666.

10 Hilton, J.R. 1983. The influence of light on the germination of *Senecio vulgaris. New Phytol.*, **94**.

11 Yaniv, Z. and Mancinelli, A.L. 1968. Phytochrome and seed germination. IV. Action of light sources with different spectral energy distribution on the germination of tomato seeds. *Plant Physiol.*, **43**, 117–120.

12 Frankland, B. 1981. Germination in the shade. In: Smith, H. (ed.), *Plants and the Daylight Spectrum*. London New York, Academic Press, 187–204.

13 Hilton, J.R. 1982. An unusual effect of far-red absorbing form of phytochrome: Photoinhibition of seed germination in *Bromus sterilis. Planta.*, **155**, 524–528.

14 Hilton, J.R. 1984. The influence of dry storage temperature on the response of *Bromus sterilis* seeds to light. *New Phytol.*, **98**, 129–134.

15 Hilton, J.R. 1987. Photoregulation of germination in freshly-harvested and dried seeds of *Bromus sterilis. J. Exp. Bot.*, **38**, 286–292.

16 Ellis, R.H., Hong, T.D. and Roberts, E.H. 1986. The response of *Bromus sterilis* and *Bromus mollis* to white light of varying photon flux density and photoperiod. *New Phytol.*, **104**, 485–496.

17 Yokohama, Y. 1965. Analytical studies on the variation of light dependence in light-germinating seeds. *Bot. Mag.*, **78**, 452–460.

18 Eldabh, R., Frederiq, H., Maton, J. and deGreef, J. 1974. Photophysiology of *Kalanchoe* seed germination. I. Inter-relationship between photoperiod and terminal far-red light. *Physiol. Plant*, **30**, 185–191.

19 Borthwick, H.A., Toole, E.H. and Toole V.K. 1964. Phytochrome control of *Pawlonia* seed germination. *Isr. J. Bot.*, **13**, 122–123.

20 Frankland, B. 1976. Phytochrome control of seed germination in relation to the light environment. In Smith, H. (ed.), *Light and Plant Development*. London, Butterworths, 447–491.

21 Kendrick, R.E. and Frankland, B. 1969. Photocontrol of germination in *Amaranthus caudatus. Planta.*, **86**, 326–339.

22 Bartley, M.R. and Frankland, B. 1982. Analysis of the dual role of phytochrome in the photoinhibition of seed germination. *Nature*, **300**, 750–752.

23 Frankland, B. 1986. Perception of light quality. In: Kendrick, R.E. and

Kronenberg, G.H.M. (eds.), *Photomorphogenesis in Plant*. Dordrecht, Martinus Nighoff, 219–235.

24 Gorski, T. and Gorska, K. 1979. Inhibitory effects of full day light on the germination of *Lactuca sativa*. *Planta.*, **144**, 121–124.

25 Gwynn, D. and Scheibe, J. 1972. An action spectrum in the blue for inhibition of lettuce seed. *Planta.*, **106**, 247–257.

26 Hartmann, K.M. 1966. A general hypothesis to interpret high energy phenomena of photomorphogenesis on the basis of phytochrome. *Photochem. Photobiol.*, **5**, 349–366.

27 Schafer, E. 1975. A new approach to explain the 'high irradiance response' of photomorphogenesis on the basis of phytochrome. *J. Math. Biol.*, **1**, 9–15.

28 Malcoste, R., Tzanni, H., Jaques, R. and Rollin, P. 1972. The influence of blue light on dark germinating seeds of *Nemophila insignis*. *Planta.*, **103**, 24–34.

29 Lona, F. 1947. L'influenza delle condizioniesterne durante l'embiogenesi in *Chenopodium amaranticolor* Costa et Reyn. Sulle qualita germinative dei semi e sul vigore delle plantule che ne derivano. *Lav. Bot.*, 324–352.

30 Pourrat, Y. and Jaques, R. 1975. The influence of photoperiodic conditions received by the mother plant of morphological and physiological characteristics of *Chenopodium polyspernum*. *Plant Sci. Lett.*, **4**, 273–279.

31 Hayes, R.G. and Klein, W.H. 1974. Spectral quality influence of light during development of *Arabidopsis thaliana* plants in regulating seed germination. *Plant Cell Physiol.*, 15, 643–653.

32 McCullough, J.M. and Shropshire, W. 1970. Physiological predetermination of germination responses in *Arabidopsis thaliana*. *Plant Cell Physiol.*, **11**, 139–148.

33 Wooley, J.T. and Stroller, E.W. 1978. Light penetration and light induced seed germination. *Plant Physiol.*, **61**, 597–600.

34 Frankland, B. 1981. Germination in shade. In Smith, H. (ed.), *Plants and the Daylight Spectrum*. London, Academic Press, 187–204.

35 Brenchley, W.E. and Warrington, K. 1931. The weed populations of arable soil. I. Numerical estimations of variable seeds and observations on their natural dormancy. *J. Ecol.*, **18**, 235–272.

36 Brenchely, H.A. and Warrington, K. 1933. The weed seed population of arable soil. II. Influence of crop, soil and methods of cultivation upon the relative abundance of viable seed. *J. Ecol.*, **21**, 103–127.

37 Milton, W.E.J. 1936. Buried, viable seed of enclosed and unenclosed hill land. *Bull. Welsh Pl. Breed. Stn.*, **14**, 55–84.

38 Wesson, G. and Wareing, P.F. 1969. The role of light in the germination of naturally occurring populations of buried seeds. *J. Exp. Bot.*, **20**, 402–413.

39 Karssen, C.M. 1980/81a. Environmental conditions and endogenous mechanisms involved in secondary dormancy of seeds. *Isr. J. Bot.*, **29**, 45–64.

40 Karssen, C.M. 1980/81b. Patterns of dormancy during burial of seeds in soil. *Isr. J. Bot.*, **29**, 65–73.

41 Roberts, H.A. and Lockett, P. 1978. Seed dormancy and periodicity of seedling emergence in *Veronica hederifolia*. *Weed Res.*, **18**, 41–48.

42 Pons, T.L. 1983. Significance of inhibition of seed germination under leaf canopy in ash coppice. *Plant Cell Environ.*, **6**, 385–392.

43 Pons, T.L. 1984. Possible significance of changes in the light requirement of *Cirsium palustre* seeds after dispersal in ash coppice. *Plant Cell Environ.*, **7**, 263–268.

44 Frankland, B. and Poo, W.K. 1980. Phytochrome control of seed germination in relation to shading. In: DeGreef, J. (ed.), *Photoreceptors and Plant Development*. Antwerpen, Antwerpen Univ., 357–366.

45 Bliss, D. and Smith, H. 1985. Penetration of light through soil and its role in control of seed germination. *Plant Cell Environ.*, **8**, 475–483.

46 van der Meijden, E, van der Waals-Kooi. 1979. The population ecology of *Senecio jacobea* in a sand dune system. I. Reproductive strategy and the biennial habit. *J. Ecol.*, **67**, 131–153.

47 Tester, M. and Morris, C. 1987. The penetration of light through soil. *Plant Cell Environ.*, 281–286.

48 Gorski, T., Gorska, K. and Nowicki, J. 1977. Germination of seeds of various herbaceous species under leaf canopy. *Flora*, **166**, 249–259.

49 Bewley, J.D. and Black, M. 1982. *Physiology and Biochemistry of seeds Vol. II. Viability, Dormancy and Environmental Control.* Berlin Heidelberg New York: Springer-Verlag: 136.

50 Pons, T.L. 1986. Response of *Plantago major* seeds to red/far-red ratio as influenced by the environmental factors. *Physiol. Plant*, **68**, 252–258.

51 Holmes, M.G. and Smith, H. 1977. The function of phytochrome in the natural environment. II. The influence of vegetation canopies on the spectral energy distribution of natural daylight. *Photchem. Photobiol.*, **25**, 539–545.

52 Smith, H. 1982. Light quality, photoperception and plant strategy. *Ann. Rev. Plant Physiol.*, **33**, 481–518.

53 Karssen, C.M. 1970. The light promoted germination of the seeds of *Chenopodium album*. VI. Pfr requirement during different stages of the germination process. *Acta Bot. Neerl.*, **19**, 297–312.

54 Duke, S.O., Naylor, A.W. and Wickliff, J.L. 1977. Phytochrome control of longitudinal growth and phytochrome synthesis in maize seedlings. *Physiol. Plant*, **40**, 59–68.

55 Bewley, J.D., Black, M. and Negbi, M. 1967. Immediate action of phytochrome in light-stimulated lettuce seed. *Nature*, **215**, 648–649.

56 Borthwick, H.A., Hendricks, S.B., Toole, E.H. and Toole, V.K. 1954. Action of light on lettuce seed germination. *Bot. Gaz.*, **115**, 205–225.

57 Vazquez-Yanes, C. and Smith, H. 1982. Phytochrome control of seed germination in the tropical rain forest pioneer trees *Cecropia obtusifolia* and *Piper auritum* and its ecological significance. *New Phytol.*, **92**, 477–485.

58 Orlandini, M. and Malocoste, R. 1972. Etude du phytochrome des graines de *Pinus nigra* arn par spectrophotometrie bichromatique *in vivo*. *Planta.*, **105**, 310–316.

59 Taylorson, R.B. and Hendricks, S.B. 1969. Action of phytochrome during pre-chilling of *Amaranthus retrofluexus* seeds. *Plant Physiol.*, **44**, 821–825.

60 von Liebenberg, A. 1884. Das stimulieren der *Poa annua* zur Keimung der samen bei Aussetzung standig wechseluder temperaturen. *Bot. Zentralbl.*, **18**, 21–26.

61 Steinbauer, G.P. and Grigsby, B. 1957. Interaction of temperature, light and moistening agent in the germination of weed seeds. *Weeds*, **5**, 175–182.

62 Toole, E.H., Toole, V.K., Borthwick, H.A. and Hendricks, S.B. 1957. *Proc. Int. Seed Test Assoc.*, 1–9.

63 Bewley, J.D. and Black, M. 1982. *Physiology and Biochemistry of seeds Vol. II. Viability, Dormancy and Environmental Control.* Berlin Heidelberg New York: Springer-Verlag: 165.

64 Thompson, P.A. 1969. Germination of *Lycopus europaeus* in response to fluctuating temperatures and light. *J. Exp. Bot.*, **20**, 1–11.

65 Todderdell, S. and Roberts, E.H. 1980. Characteristics of alternating temperatures which stimulate loss of dormancy in seeds of *Rumex obtusifolius* and *Rumex crispus*. *Plant Cell Environ.*, **3**, 3–12.

66 Probert, R.J., Smith, R.D. and Birch, P. 1986. Germination responses to light and alternating temperatures in European populations of *Dactylis glommerata*. V. The principal components of the alternating temperature requirements. *New Phytol.*, **102**, 133–142.

67 Probert, R.J., Gajjar, K.H. and Haslam, I.K. 1987. The interactive effects of

phytochrome, nitrate and thiourea on the germination responses to alternating temperatures in seeds of *Rannunculus sceleratus*. A quantal approach. *J. Exp. Bot.*, **38**, 1012–1025.

68   Goudey, J.S., Hargurdeep, S.S. and Spencer, M.S. 1987. Seed germination of wild mustard (*Sinapis arvensis*): factors required to break primary dormancy. *Can. J. Bot.*, **65**, 849–852.

5
Seedling development

Patterns of seedling development

The process of seed germination ends, by convention, with the beginning of elongation of the cells of the embryonic axis. The subsequent processes are part of seedling development. In dicotyledons the pattern of development may be epigeal or hypogeal. Generalised diagrams of these two patterns may be seen in Figure 5.1. The significant difference between these two types is the form and function of the hypocotyl. Hypogeal seedlings such as pea (*Pisum sativum*) and broad bean (*Vicia faba*) have a rudimentary hypocotyl that does not elongate. The aerial parts originate from the plumule or terminal bud of the epicotyl. Epigeal seedlings such as french bean (*Phaseolus vulgaris*) and cucumber (*Cucumis sativus*) have hypocotyls which elongate, thus raising the cotyledons above the level of the soil where they frequently develop photosynthetic competence but usually atrophy soon after the first foliage leaves have expanded. The apical meristems of the shoots are protected by a plumular hook in hypogeal development and a hypocotylar hook in epigeal germination. It is envisaged that the broad surface of a hook pushing through the soil particles reduces abrasive damage that might otherwise occur to the more limited surface area of the apical meristem. The plumular hook of a pea is the site of elevated ethylene production. A short period of red light causes a transient decrease in the production of ethylene (see p. 72) and as a result the hook opens.[1] If ethylene is exogenously supplied, the hook will re-close even in the light. In the natural environment, the hook would receive the appropriate light treatment as it emerges from the soil. Similar results were reported for beans' hypocotyl hook.[2]

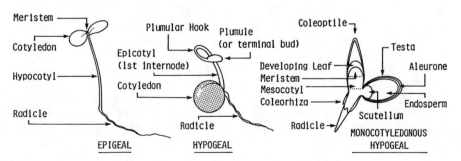

Figure 5.1 Generalised patterns of seedling development.

Cereal seedlings are generalised in Figure 5.1 and have a hypogeal pattern of development. The single cotyledon, being modified to form the scutellum, remains at or below the soil surface. The meristem and young developing leaves are protected during early development by a sheath of tissue called the coleoptile. This is a modified leaf which splits after a few days, allowing the leafy shoot to emerge. The mesocotyl, which subtends the apex and the young leaves, may also be referred to as the first internode. The radicle is also protected in early development by a sheath known as the coleorhiza; this splits and allows the primary root to emerge. Variations on these generalised patterns of development are known and described elsewhere.[3]

Light within the hypocotyl and the coleoptile

The propagation of light down a plant organ has been compared with the behaviour of light in a fibre optic strand. When light has entered a fibre optic strand, which it does over a limited acceptance angle, it travels down the fibre via multiple reflections from the wall of the strand. Organs such as hypocotyls, coleoptiles and mesocotyls, which add to the elevation of the seedling, possess files of cells which have a longitudinal axis 3 or 4 times greater than their diameter. The efficiency of a light reflectance system which contains cross walls would be much lower than that of a homogeneous system like a fibre optic strand. Light does pass through cytoplasm, vacuoles and cross walls, but the intercellular spaces between the longitudinal cell walls and the longitudinal cell walls themselves may conduct more efficiently. The water content of the cell walls may enhance the efficiency of light conductance as may forward scattering of light by various entities. Using etiolated mung bean hypocotyls, which have an acceptance angle of 59°, Mandoli and Briggs[4,5] were able to show that over short distances a coherent image can be transmitted and that over a distance of 4.5 cm etiolated tissue has about 1 per cent the efficiency of a fibre optic strand. Despite this low efficiency, such transmission of light down etiolated organs newly emerged from the soil can have radical effects upon the subsequent development of the seedling (see below).

Measurement of light within the bean hypocotyl presents a simple picture compared with similar measurement within the coleoptile. The hypocotyl can be regarded as a homogeneous cylinder compared with the variable anatomy of the coleoptile. The apex is entire tissue for about 12 to 15 cells. In the mid-region it becomes a hollow cylinder and in the basal region it surrounds a number of primary leaves. The study of the absorption of light by the coleoptile of *Zea mays* has led to an understanding of the light gradients created in experiments on phototropism.[6] The absorption of unilaterally supplied B, 450 nm, is almost complete at a depth of 0.5 mm in the region of the apex, whereas in the mid-region the boundary between the inner epidermis and the hollow centre of the coleoptile allows transmission of very similar levels of energy to the inner epidermis of the unilluminated side. Absorption continues as the light penetrates the unilluminated side. Recent work has indicated that phototropic responses depend on the measurement of light gradients across the coleoptile[7] (see also p. 39). Such hypotheses may need to consider that in this region the B fluence is very similar at the inner epidermis on either side of the hollow but very different at the illuminated and unilluminated outer epidermis. By contrast, B entering the basal region of the coleoptile is largely absorbed, first by the coleoptile sheath and then by the primary leaves. However, a measurable quantity of light is scattered within the hollow of the coleoptile and is absorbed by the inner tissues of the unilluminated side of the coleoptile.

The Thomson hypothesis

Working with *Avena* seedlings, Thomson (1950)[8] showed that the response to light of a plant organ depended on the stage of development of the organ. He suggested that light accelerates all phases of development, division, elongation and maturation. Thus, young organs may move rapidly into the elongation phase whereas older organs may move into maturation without achieving their full growth potential. In consequence an organ, particularly a coleoptile, may increase in length in response to light at the early stage of growth but the same species and the same organ can be inhibited in its growth by light at a later stage.

Control of mesocotyl and coleoptile development

The Graminae are a family of plants which have been extensively used experimentally because of the economic importance of cereals, and photomorphogenic studies are no exception to this rule. Studies of the effects of light on cereal seedlings have fallen into two main categories: the effects of phytochrome on elongation growth and the effects of blue light on phototropic growth (p. 34).

Phytochrome has been shown to control promotion of coleoptile growth and inhibition of mesocotyl growth in *Avena*[9] and *Zea*[10] although the reversibility of oat coleoptile growth was not demonstrated until more recently.[11] *Triticum* and *Hordeum* display phytochrome control of coleoptile elongation in excised sections[12] whereas in intact seedlings, phytochrome has been shown to alter the pattern of development rather than the overall length.[13] *Oryza sativa* has a coleoptile which can be inhibited by R and the effect reversed by FR.

A number of factors has added to the difficulties of studying coleoptile growth. First among these has been the practice of treating imbibed cereal seeds with dim red light to inhibit mesocotyl growth while producing 'good' coleoptiles for experimentation. This literature is now difficult to interpret in the light of present knowledge (see VLFR below). Similarly, work with sections of coleoptiles versus intact seedlings and comparisons of sections with apices to sections without apices have also caused confusion. The work considered here results primarily from experiments with intact seedlings. In accordance with the Thomson hypothesis, the effects of illumination on coleoptiles vary with the development stage of the coleoptile. Phytochrome can affect development of coleoptiles of *Triticum* seedlings in a number of ways[14] Coleoptile growth is promoted in the tip region but is inhibited in the basal region. The cells of the middle region of the coleoptile show both promotion and inhibition and the effects cancel one another out, as indeed they can when the entire coleoptile is measured. In a detailed study of wheat cultivars[13] it was found that all had a common pattern of growth. The apical 5 mm zone is stimulated by R whereas the 5–10 mm zone is only slightly affected and growth in the 10–15, 15–20 and 20–25 mm regions is inhibited. The overall effect of R on coleoptiles of intact seedlings of wheat and barley varies with the age of the seedling. In young seedlings with 20- to 29-mm coleoptiles, the R treated seedlings and the dark controls achieve the same length after 24-h of growth. With older seedlings the extension of cells in the upper part of the coleoptile was of less influence on overall growth than the inhibitory effects on the elongation on the basal cells of the coleoptiles. In consequence, the R-treated plants are shorter than the dark controls after a 24 h growth period. These findings are not necessarily contradictory to the findings of others[15] who used a LVDT (see p. 72 and Figure 5.3) and demonstrated a R-induced inhibition of wheat coleoptile extension, beginning 10 min after the onset of radiation. The inhibition is transient. The inhibition caused by 1 min of R recovers to the dark control rate in 2 h, 5 min of R in 3 h and continuous R (by extrapolation of the data) in approximately 4 h. This response is not FR-reversible. Lawson and Weintraub suggested that the coleoptile growth is controlled in terms of independent action of light and IAA on microtubules.[16] The microtubule was assigned the role of controlling microfibril deposition in the cell wall, and these workers suggest that light and auxin each stimulate different compartments of the microtubule subunit, causing increased polymerisation and increased growth.

The tip of the rice coleoptile is generously supplied with phytochrome and the elongation of this organ can be inhibited by R and the effect reversed by

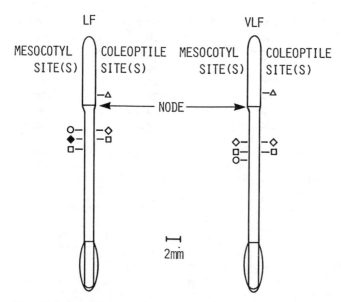

Figure 5.2 Location of phytochrome which influences mesocotyl and coleoptile development in etiolated oat via LFR and VLFR. Different symbols relate to small differences in the method by which the sites of photoperception were detected (after Mandoli and Briggs, 1982).

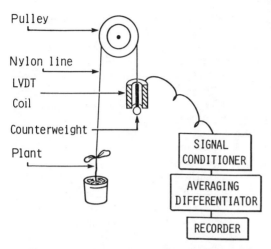

Figure 5.3 Linear variable displacement transducer.

FR.[17,18] While the coleoptile tip is the most sensitive region for light detection and will emerge into the light first, the region of cell elongation is found midway down the coleoptile. Evidence exists[19] to suggest that the presence of Pfr can inhibit polar transport of IAA from the tip to the region of elongation and thereby inhibit growth. However, the inhibition of polar IAA transport is transitory and the hypothesis needs to be expanded to suggest that, once starved of IAA, the sensitivity of the elongation zone to auxin is decreased.

Probably the best understanding of seedling development comes from studies of oat. Irradiation of dark grown *Avena sativa* seedlings with a brief period of red light stimulates the elongation of the coleoptile and inhibits the elongation of the mesocotyl.[20,21] Mandoli and Briggs showed that each of these growth responses is biphasic[22] and the phases are now known as the very low fluence response (VLFR) and the low fluence response (LFR). An example of these biphasic responses was previously shown in Figure 3.5. The presence of the VLFR has necessitated the growing and manipulation of experimental material in complete darkness. These workers went on to find that the sensitivity to red light reached a maximum in the region of the coleoptile node.[23] By exposing only 2 mm sections of the intact coleoptile or the mesocotyl to light they were able to locate the points of perception for these responses (Figure 5.2). Since the sites of perception are separate from the sites of elongation, they suggested the seedlings had light-piping qualities. By quantifying the fibre optic properties of the etiolated tissue they were able to predict the degree of growth response of either the coleoptile or mesocotyl. Thus the emerging coleoptile tip is able to pipe light down through the soil to the sites of response beneath, enabling the developmental changes to be initiated in parts of the plant not directly illuminated.

The elongation of the coleoptile is a system in which photoreceptor/hormone interaction has been examined.[24] By removing the tip of the coleoptile of *Avena*, it has been shown shown that the light sensitivity of the section can be greatly enhanced ($\times 10\,000$) by the presence of 6 μM IAA. It may therefore be concluded that in *Avena*, at least, light and IAA are both capable of controlling coleoptile and mesocotyl growth and are capable of interacting. The interaction of light piping and IAA is still a matter of speculation.

The control of the mesocotyl in *Avena* has been investigated together with the coleoptile[23] while in other species it has been investigated separately. R inhibits mesocotyl development in *Zea mays*.[25,21] These results complement the findings with oat in as much as the VLFR and the LFR are present and coleoptile promotion and mesocotyl inhibition are recorded. The results of various workers agree with the hypothesis of Van Overbeek (1936)[26] that light via phytochrome regulates mesocotyl elongation by regulating the supply of IAA from the coleoptile. For example, when treated with R the growth rate of the maize mesocotyl falls to 20 per cent of the dark control in 4 h. Over the same period of time the IAA concentration in the mesocotyl falls by 50 per cent.[27] The growth rate recovers after 10 h, as does the auxin supply. However, data exist which are not in accord with the Van Overbeek hypothesis. R inhibits mesocotyl elongation in seedlings from which the coleoptile has been removed[28,29] and this inhibition cannot be relieved by exogenously supplied IAA.[27] It may be argued that since the mesocotyl is itself a site of photoperception[23] the hormonal communication with the coleoptile tip is redundant. If both concepts are correct then, in the natural situation, the coleoptile would emerge into the light first, the IAA supply to the mesocotyl would be reduced and cell elongation retarded. When the photoperceptive sites of the mesocotyl emerge into the light, the control of mesocotyl development may pass from the coleoptile directly to this organ.

In an investigation of changes in the cell plastic extensibility in the mesocotyl region of the *Zea mays* seedlings, it was found that 3 h of WL inhibited elongation and caused a sharp decrease in plastic extensibility while elastic extensibility remained unaffected.[30] If the seedlings received FR at the end of the WL treatment, after 20 h in darkness their growth recovered to 90 per cent of the dark control. This effect was FR/R reversible. Plastic extensibility made only a partial recovery after 20 h whereas elastic extensibility remained unchanged under all treatments. It was reported that a range of IAA concentrations was unable to reverse the inhibition imposed by light whether intact seedlings or decapitated coleoptiles were used. The evidence of this work indicates that IAA and phytochrome act via separate pathways upon mesocotyl growth in *Zea*.

BAP and straight growth in cereal seedlings

The effect of B on phototropic growth of coleoptiles has been extensively researched but much less is known about the effects of B on coleoptile elongation growth. B is known to reduce growth in the basal region of the wheat coleoptile while promoting growth in the apical region.[14] These effects were very similar to the effects of R and it was shown that the B effects could be reversed by FR irradiations. Only the finer points of the data cannot be explained in terms of phytochrome but the involvement of BAP cannot be ruled out.

High fluence rates of B are required to demonstrate BAP responses in dicotyledons,[31,32] and light sources which are capable of producing such fluence rates of B have only become commonly available in recent years. Negative findings in the older literature often lose credibility since the absence of a response may be due to the use of a B fluence below the threshold of reception for BAP. Action spectra for phototropism and photoinhibition of straight growth in *Avena* coleoptiles are very similar.[33] Both action spectra have maxima at 360, 440 and 470 nm. This is a strong indication that the BAP is involved not only in phototropism (see p. 37) but also in the control of straight coleoptile growth. The high fluence of B in the natural environment is capable of stimulating BAP and has been shown to be important in controlling juvenile structures in dicotyledons which were previously thought to be solely controlled by phytochrome.[34,32] The suggestion that BAP is ubiquitously involved in the control of straight growth of coleoptiles in the natural environment would be speculative since the majority of reports refer to etiolated tissue and there have been few studies of coleoptile growth in green tissue or during the transition between these two forms. However, experiments have been done with *Zea mays* seedlings grown in darkness for 21 h and then given 8 h of WL per 16 h of D for three cycles.[35] Whereas this represents a laudable attempt to simulate natural conditions it must be pointed out that the fluorescent light used was 5 per cent that found on a clear summer day at temperate latitude. End-of-day FR treatment increased the rate of mesocotyl and coleoptile growth, the presence of chlorophyll not appearing to interfere with the perception of this signal. The coleoptile response was the opposite of that obtained by other workers.[21] Since this also occurs in the presence of Sandos 9789 (otherwise known as Norflurazon), a chlorosis-inducing herbicide, it may be deduced that chlorophyll and photosynthesis play no part in this response.

The role of the coleoptile apex

Since the time of Went (1928)[36] it has been generally accepted that the apex secretes IAA, which diffuses basipetally. This concept arises largely from decapitation experiments in which the growth of the coleoptile is monitored after the apex has been removed or when after decapitation IAA sources are replaced in various ways. These tenets have been re-examined using modern techniques which are more discerning both in spatial and temporal terms.[37,38] Using coleoptiles from *Zea* and *Avena*, Parsons and co-workers confirm that when coleoptiles are decapitated the growth rate is markedly reduced. However, the growth rate reduction resulting from decapitation does not occur over a time period nor in a pattern which could be explained in terms of basipetal transport of IAA from the apex. When coleoptiles are decapitated, the remaining segment recovers from the initial reduction in growth rate and this has always been associated with the regeneration of a 'physiological apex'. Once again the zone where this growth recovery occurs is a zone which could not be affected by the re-establishment of basipetal transport of IAA. These workers point out that the concentration of exogenously supplied auxin required to prevent the decrease in growth rate is far above physiological concentrations and that if a band of 2-chloro-9-hydroxyfluorene-9-carboxylic acid methyl ester(CFM), a morphactin which is an efficient inhibitor of basipetal IAA transport, is applied in lanolin to the outside of the coleoptile, the growth rate is not significantly altered. It would appear from this evidence that the traditional notion that IAA secreted from the apex arrives in rate-limiting quantities in the region of cell elongation is under threat. In consequence, also in danger would be those theories which would link the phytochrome control of coleoptile development to the basipetal transport of IAA.

Phytochrome control of ethylene biosynthesis in etiolated seedlings

In recent years, as suitable assay techniques have become available, attention has turned to the interaction of phytochrome and ethylene. However, all reports in the literature concern etiolated material and, in consequence, extrapolation of these effects to the natural environment must be limited to emergence from the soil. As has been mentioned earlier, ethylene production has been shown to be reduced by light in the plumular hook portion of a pea,[1] and similar effects were reported in beans[2] where the light reduction of ethylene level was implicated in the control of the opening of the hypocotylar hook. The reduction of ethylene levels was shown to be R/FR reversible in *Oryza sativa* coleoptiles[39] and similar effects are reported for soybean (*Glycine max*) hypocotyls where ethylene production was reduced by 45 per cent by R irradiation.[40] The reduction of ethylene levels by Pfr may not be a ubiquitous effect, as work with *Sinapis alba*,[41] the cotyledons of *Cucumis sativus*[42] and pea epicotyls[43] shows that light increases the level of ethylene production while no effect of R was found in *Lactuca sativa*.[44] A degree of elucidation is supplied by work with beans[45] where phytochrome reduces the level of ethylene production, but the effect is age-dependent and cannot be detected in beans younger than 6 or older than 11 days.

Linear variable displacement transducer (LVDT)

First used for investigating plant growth by Meijer (1968),[46] the LVDT has proved to be a very useful tool in revealing the small and sometimes transient changes in growth rate induced by changes in the light environment. The LVDT comprises a set of specially tapered primary and secondary coils through which a metal cylinder or core is able to pass. The LVDT may be used in several ways to measure plant growth and a simple one is shown in Figure 5.3. The plant is attached to a fine nylon line which runs over a pulley and is attached to the core. As the plant increases in length, the nylon line, which is kept under tension by the counter weight, lowers the core through the coil. This causes the output of the coil to change and the signal is differentiated to produce a voltage proportional to the elongation rate of the plant and may be recorded either by microcomputer or by chart recorder.

Cell elongation

The use of LVDT has revealed that plants grow and cease to grow within a few minutes of the appropriate light stimulation. Growth in these studies does not refer to an increase in dry weight of the plant but rather an increase in the length of cells. In a recent review, Cosgrove (1987)[47] refers to the process of cell growth as a 'push-pull' system. The cell wall has two mechanical components which oppose the stress exerted on it by the turgor pressure of the protoplast. The first consists of elastic elements which are the macromolecules, such as cellulose, which form networks in the cell wall and these are able to stretch under tension and return to their original shape when tension is relieved. The second component of cell walls is the non-elastic or plastic elements. These elements shear or distend irreversibly under tension. Growth starts when wall yielding reduces the wall stress, and this is thought to lower the water potential of the protoplast by 0.3 to 0.5 bars. During this process the plastic elements are thought to lengthen, although at this stage there is no increase in cell length. As plastic lengthening occurs, the elastic elements relax and return to their original shape, which indicates that the wall pressure has been released. As was indicated above, the protoplast would then be able to take up water from its surroundings as a result of a decrease in the water potential. The resulting influx of water causes the cell to grow via a process known as wall creep. It is not known how photoreceptors are able to accelerate or decelerate these processes.

Leaf unrolling in cereals

Leaves of cereals emerge from their protective sheath where they have developed in a rolled condition. In laboratory experiments where 7-day-old dark-grown seedlings are used, it has long been known that phytochrome controls leaf unrolling.[48] Despite the artificial conditions of these experiments and the lack of experimentation under natural or simulated natural conditions, there is no reason to suppose that this is not an indication of what occurs in the natural environment. The mechanism of this response is still unclear but apparently humidity plays a part in the differential expansion of cells and phytochrome and gibberellins interact to control this response in etiolated material.[49] Although transcription does not seem to be involved in early stages of leaf unrolling,[50] the inhibition of transcription by actinomycin later inhibits leaf unrolling. Various workers[51] have indicated that Ca^{2+} is involved in this response, perhaps as regulator of transport between cells. It has been suggested that Ca^{2+} is involved in two processes, one of which must be completed within the first hour of the onset of irradiation.[52]

Control of hypocotyl development

The role of BAP in the control of straight growth of the mesocotyl or coleoptile is far from established. By contrast, the involvement of BAP in the control of hypocotyl elongation has long been known.

Meijer[53] reports that the hypocotyl extension of *Cucumis sativus* var. Venlo Picklin in etiolated tissue could be inhibited by B but not by R or green. Later[46] he showed that by using a LVDT to measure growth (see above) the inhibition of hypocotyl elongation could be detected after 60 s with B, whereas if R was used, a much longer period was required (15 to 90 mins). Black and Shuttleworth (1974)[54] found that hypocotyl elongation of de-etiolated *C. sativus* var. Ridge Greenline could be inhibited by both B and R. In experiments where the cotyledons were covered with aluminium foil, these workers were able to show that the effect of R on hypocotyl elongation was exerted by phytochrome and the site of perception was primarily the cotyledons. On the other hand, the inhibition caused by B was detected by the hypocotyl, the cotyledons being insensitive to B. LVDT studies of Gaba and Black using this variety of cucumber[55] showed that when light-grown plants were transferred from dark to light, hypocotyl elongation was inhibited by B within 5 min, whereas R caused inhibition after 5 h. When plants were transferred from light to dark they found that the R-induced inhibition persisted for 8 to 10 h whereas B inhibition disappeared after about 20 mins. Although evidence exists to suggest that phytochrome and BAP interact in some way to increase the sensitivity to BAP at low B fluence rates,[56] this does not detract from the main findings of the above workers that phytochrome and BAP control hypocotyl growth independently. The interpretation of these data in terms of the natural environment is not easy. At low fluence rates at dawn, the synergistic interaction of BAP and phytochrome could be of importance, whereas the high Pfr levels which might be engendered on clear nights,[57] coupled with the persistent inhibition of hypocotyl extension by phytochrome on short summer nights, would indicate that hypocotyl growth in the natural environment proceeds at very low or inhibited rates. If the natural habitat of *C. sativus* is grassland, competing only with a low level canopy, then the function of the hypocotyl would be to respond to vegetational shade by extension growth, elevating the aerial parts above the surrounding canopy.

The involvement of BAP in hypocotyl control is not unique to *Cucumis*. Work not only with this species but with light-grown *Latuca* and *Lycopersicon* using the SOX technique (see p. 36) shows that high levels of Pfr are capable of inhibiting hypocotyl extension but additional inhibition is apparent when blue light is added.[58] Thomas *et al.* (1980)[59] go on to demonstrate a temporal difference between the effects of B and FR on lettuce seed hypocotyl inhibition. The FR effect is seen only in young seedlings and lasts for only 24 h and is removed by a short red pre-treatment. In contrast, B is highly inhibitory at any age, and the effect of B persists throughout illumination and is not abolished by a red pre-treatment.

A great deal has been learned from the hypocotyl of *Sinapis alba* about the mechanism of phytochrome action. However, it was maintained for some years[60] that phytochrome was the sole receptor responsible for the control of hypocotyl elongation in this system. The involvement of BAP with this response was first recognised as a result of LVDT work by Cosgrove (1982)[34] and finally confirmed and accepted by Drumm-Herrel and Mohr (1985).[32] The LVDT technique showed that blue light inhibited hypocotyls of etiolated *S. alba* with a 40- to 60-s lag phase, whereas R and FR have slower and weaker effects. These findings were confirmed by others[61] who, using etiolated *Sinapis alba*, were able to show that hypocotyl growth was inhibited by R after a lag phase of about 5 min, whereas B acts with a lag of only 1 min. In etiolated seedlings, at least, B inhibition can be demonstrated in the first 5 days of seedling development only.[32] After this period, when etiolated seedlings are exposed to light, the inhibition of hypocotyl growth is controlled by phytochrome alone.

There may be a selective advantage in the hypocotyl of *S. alba* being sensitive to both B and R : FR. It would enable the young seedling to detect neutral density shade, supplied perhaps by an uneven soil surface during early emergence and shade detection seen afterwards.

The hypocotyls of *Helianthus annuus* and *Cucumis sativus* respond to B with lag phases of 60 and 30 s, respectively. It is difficult to explain such rapid responses in terms of BAP affecting hormone supply to elongating cells or indeed changes of cell sensitivity to hormone, and a hypothesis which involves cell wall metabolism is preferred.[62] Continuing this theme, it was found that B inhibits hypocotyl elongation in *H. annuus* and *C. sativus* by decreasing the yielding properties of cell walls. These properties change very rapidly after a lag period and the change (but not the lag period) is a function of fluence rate.[63]

Development of the cotyledon

Dicotyledonous seedlings which exhibit epigeal development depend in part upon the cotyledons developing photosynthetic competence until the first leaves are capable of performing this function. At this stage, or shortly afterwards, the cotyledons atrophy.

If the cotyledons of mustard are detached from the embryonic axis at an early stage of germination they are able to form more chlorphyll but develop less photosynthetic capacity than cotyledons which remain attached to the hypocotyl.[64] Oelze-Karow and Mohr[65,66,67] showed that early changes in the development of the cotyledon depend on the rapid transmission of signals after Pfr formation has occurred in the hypocotyl hook. Although little is known about this signal, a biophysical signal is favoured over a biochemical signal because of the rapidity with which it is transmitted. This signal is an absolute requirement for the development of photophosphorylating ability in the cotyledon although significant amounts of chlorophyll are formed in its absence. The initiation of the signal is a threshold phenomenon and the Pfr in the hypocotyl must exceed 1.5 per cent. The synthesis of chlorophyll is not independent of the hypocotyl in all species. The cotyledons of *Cucumis sativus*, for example, depend upon the presence of the hypocotyl for chlorophyll synthesis.[68] In mustard seedlings grown at 25°, cotyledon expansion takes place at about 36 h. Although this is a phytochrome-mediated phenomenon the response escapes reversibility 15 h after imbibition. The time lag between loss of photoreversibility and expression indicates the longevity of some of the components of the phytochrome-triggered signal chain.[69] It was also reported that the light-mediated cotyledonary expansion results from an increase in cell wall extensibility.

Cotyledon as a light trap

The green cotyledon of the hypogeal seedling *Cucurbita* has been extensively used as a model system for understanding the absorption of light for both developmental and photosynthetic purposes. The *Cucurbita* cotyledon has distinct layers of palisade and spongy mesophyll and this allows a degree of extrapolation from the results found with this tissue to the behaviour of light within a dicotyledonous leaf. The behaviour of light entering a green leaf is discussed later (p. 84).

Light and chlorophyll gradients within green cotyledons of *Cucurbita pepo* have been measured by inserting fibre-optic microprobes.[70] Half of the chlorophyll was found to occur in the adaxial 300 μm of the cotyledon, which was 1200 μm thick. When the adaxial surface of the cotyledon was irradiated with collimated light almost complete absorption of 680 nm was found in the first 400 μm, within the palisade region of the cotyledon. Wavelengths such as 550 and 750 nm were scattered by the palisade cells, and the spongy mesophyll received only diffuse radiation. Scattering is also found and is more efficient in the spongy mesophyll, which is probably the result of the greater frequency of air spaces. When the abaxial surface was irradiated, 680 nm of light penetrated to a depth of 700 μm.

Effect of light on radicle and root development

In the vast majority of both monocotyledons and dicotyledons, the radicle emerges from the seed first. Subsequent growth of the radicle is inhibited by light. One function of the radicle is to provide anchorage for the development of the aerial parts of the plant. Radicles are associated with downward growth and are said to be positively gravitropic. (Gravitropism is the term which has replaced geotropism.) A large body of evidence exists to suggest that the gravity-sensing mechanism for both root and shoot resides in the root and an important part of the sensing mechanism is the presence of amyloplasts in the cells of the root cap.[71]

The roots of *Vicia faba* have been examined using electron microscopy and it was found that starch grains sedimented as a result of gravitational pull and that only a few cell particles behave in this way.[72] Evidence has now accrued which correlates amyloplast sedimentation with all major aspects of the gravitropic-sensing mechanism. The link between amyloplast sedimentation and the gravitational growth response is less certain, although a good deal of evidence suggests that ABA present in the root cap is transported laterally across the root cap under the influence of gravity, resulting in an asymmetric distribution. The ABA is then thought to be transported to the cell extension zone, where growth is asymmetrically inhibited, causing gravitropic bending. If ABA did not have a ubiquitous role in gravitropism, and other growth inhibitors were able to function in a similar way, then many inconsistencies in the literature could be explained.[73,74]

It has been known for some years[75] that significant levels of phytochrome occur in plant roots, and in etiolated oat seedlings phytochrome is localised in the root tip.[76] Furthermore, the roots of some plants, *Convolvulus arvensis* and some *Zea mays* varieties, grow in a random fashion in darkness and need an exposure to light to sensitise the roots to the gravitropic stimulus.[77] When the terminal 1 mm of roots of *Convolvulus arvensis* are exposed to short periods of red light, the roots become gravitropically sensitive. No evidence is available as to the locus of interaction between phytochrome and gravitropism.

An action spectrum of light-induced gravitropism in *Zea mays* roots shows two peaks in the blue region and a broad peak in the red region (600 to

700 nm).[78] The action of BAP or an interaction of phytochrome and BAP cannot be excluded from this response.

Some interesting results have been achieved with an agravitropic mutant of *Arabidopsis thaliana*.[79] Although the seedling of this mutant, aux-1, produces roots which do not respond to gravity, they retain the ability to respond to light. In consequence, by using this mutant in conjunction with the wild type, it is possible to investigate the effect of these two vectors separately. Using fluorescent WL at a fluence of 160 μmol m^{-2} s^{-1}, it was shown that in addition to the negative phototropic and positive gravitropic curvature recognized in roots, both WL and gravity induced horizontal coiling. While the root is on the soil surface and exposed to radiation from above, these vectors act synergistically to cause horizontal clockwise curvatures. It might be suggested that such an effect aids the seedling in initiating root burial.

Light and root growth

Light has been shown to inhibit the elongation of roots of a number of species[80] and it would appear that the root cap is the site of photoperception.[81,82] The mechanism by which light brings about this inhibition has been discussed at some length[83,84,85] and it has been suggested that light promoted the formation of a non-volatile growth inhibitor which might be ABA[86] or xanthoxin.[81,82,87] The role of ABA and similar growth inhibitors is not unanimously agreed upon and it has been suggested that ethylene may play a part in this system.[88] It has been known for some time[89] that ethylene is capable of inhibiting root growth. The substance 3,5,-diiodo-4-hydroxy-benzoic acid, which promotes growth of roots particularly in the light, is an effective inhibitor of ethylene production in cress roots, and it has been suggested that light could inhibit root growth via the stimulation of ethylene production.[88] This finding has received some support since it has been shown that ethylene was produced by *Zea* and *Pisum* roots.[90] Recently it was demonstrated that the growth of pea roots was inhibited 40 to 50 per cent by WL and there was a concomitant production of ethylene.[91] Further, the substance 1-aminocyclopropane-1-caboxyacid (ACC) was shown to stimulate dark-grown roots to produce ethylene in quantities similar to those produced by roots in the light where there was a concomitant inhibition of root growth. Inhibitors of ethylene synthesis were partly capable of reversing the growth inhibition imposed by light. It was concluded that light-stimulated ethylene production is at least partly responsible for light inhibition of root growth.

Adventitious root formation

Stem cuttings of mung beans (*Phaseolus aureus*) form adventitious roots when supplied with auxin in solution and do so more abundantly under conditions of high fluence (70 Wm^{-2}).[92] If the cuttings are placed in the dark then the plants which are the most successful are the ones grown under the highest fluence rates. On the other hand, if stem cuttings do not receive an exogenous supply of auxin, then high fluence rates are found to be inhibitory. It was also shown that high fluence enhances the uptake of auxin into the plant.

Hypocotylar and plumular hook openings.

Hypocotylar and plumular hooks have long been interpreted as strategies which protect the apical meristems from abrasive damage as the seedlings emerge from the soil. A time-lapse photographic study of the de-etiolation process of a small sample of *Pisum sativum* showed that, after exposure to R, there is a lag phase of 2 to 5 h before the plumular hook begins to open.[93] These plants reach a maximum opening rate between 8 and 11 h and opening is completed by 11 h for rapid individuals and incomplete at 17 h for plants that respond more slowly. A detailed study of the hypocotyl hook of *Phaseolus vulgaris* shows that R stimulates cell elongation on the inside of the elbow of the hook and at the top of the shank.[94] This cell elongation occurs under continuous illumination with R and causes hook opening to begin after a 2-h lag phase. Several lines of evidence were produced which indicated that R interacts with endogenous auxin to produce hook opening.[95] First, acidic indoles have been detected in hook diffusates, second, IAA inhibitors prevent hook opening in R, third, IAA stimulates elongation of shank cells and fourth, R increases tissue sensitivity to IAA.

Epicotyl (first internode) development

Although the term epicotyl is used to describe the first internode of both epigeal and hypogeal seedlings, the function of this organ during seedling development differs in the two classes. In hypogeal types, the epicotyl is closely associated with the emergence of the seedling from the soil and the subsequent straightening of the plumular hook. Seedlings which develop in an epigeal manner are capable of forming an epicotylar hook, but this occurs largely under artificial conditions of prolonged darkness. In the natural condition the epicotyl of epigeal seedlings is primarily associated with subtending the first foliage leaves, since emergence from the soil is achieved by hypocotyl elongation.

Time-lapse photography has shown that during de-etiolation of *P. sativum*, stem elongation was inhibited by red or white light after a 6 h lag phase.[93] This inhibition was transient, ending 18 h later. In another study, designed to locate the site of R perception, it was shown that when dark-grown pea seedlings were given 10 mins of light per day, after 4 days epicotyl growth decreased but increases were noted in later internodes and similarly there were increases in leaf development. Nodes were notably more sensitive to light than internodes.[96]

Photocontrol of epicotyl and subsequent internodes

Today's understanding of the photocontrol of internode growth originated with the work of Holmes and Smith.[97–100] These workers assumed that phytochrome in green plants would behave in much the same way as phytochrome from etiolated seedlings. In consequence, they measured the phi or Pfr/Ptot in etiolated tissue which had been exposed to light from various natural and artificial sources. The R : FR of these lights was then measured.

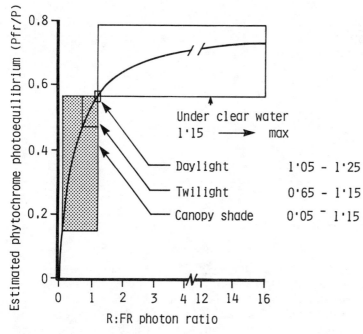

Figure 5.4 Relationship between phytochrome photoequilibrium and R : FR. The shaded area over the steepest part of the curve indicates the changes in R : FR, and consequent effects on phytochrome photoequilibrium, which occur in the natural environment (after Smith, 1982).

This value is referred to as zeta (ζ) in the earlier literature (see also p. 16). The relationship between R : FR and ϕ is reproduced in Figure 5.4.

This figure shows that phytochrome undergoes its greatest change in photoequilibrium under the influence of R : FR, which occurs in the natural terrestrial environment. These findings led Morgan and Smith (1976)[101] to grow *Chenopodium album* (Fat hen), a weed of arable land, under different R : FR, while maintaining photosynthetically active radiation (PAR) at the same level for each treatment. It is generally accepted, though not entirely certain, that FR wavelengths beyond 705 nm make no significant contribution to photosynthesis. The effect of the R : FR on the phytochrome of the plant can be estimated either by calculation[102] or by using the relationship derived by Smith and Holmes (1977).[100] At the beginning of the experiment, these plants have equal potential for photosynthesis, but the strategy engaged by these individuals varies with ϕ. At low ϕ the plants produce extended internodes and at high ϕ they produce shorter ones. After 2 weeks of treatment, the plants under simulated shade (i.e. low ϕ) achieve an overall height three times that of the plants in the simulated open situation. A straight line relationship, shown in Figure 5.5a, between log. stem extension rate and ϕ could be derived. This indicates that phytochrome directly controls the overall heights achieved.

These findings led Morgan and Smith to investigate the photocontrol of

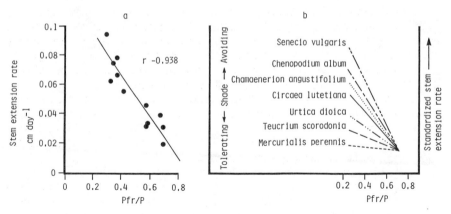

Figure 5.5 Relationship between phytochrome photoequilibrium and extension growth. (a) Detailed data for *Chenopodium album* (b) data from a range of herbaceous species.

stem growth in a number of species from a range of habitats, from open situation to heavy shade.[103] They found that low R : FR had a much greater effect on the stem extension rate of species native to open situation and shade-intolerant than on those native to either light or heavy shade. In most species the relationship between log. stem extension rate and ϕ was linear. These results, shown in Figure 5.5b, show that when a ruderal such as *Senecio jacobea* is exposed to low ϕ it is capable of making large shade avoidance responses by internode extension. On the other hand, such a strategy plays no part in the survival of a below-canopy native of the woodland habitat such as *Mercurialis perrenis*. These workers also reported effects on petiole length and leaf: stem dry weight.[104]

The majority of the work published on internode growth investigates either the first internode of a plant which has not developed much beyond this stage or measures increases in overall height. However, Bain and Attridge (1988), in their study of *Galium aparine*, paid particular attention to the influence of R : FR on the lengths of individual internodes.[105] Seed of *Galium aparine* (Cleavers), a self-fertile weed of arable land which can be particularly damaging to cereal crops, was gathered from plants growing in the hedgerows of an arable field and from plants growing in the open situation some 30 m away from the hedge. Young plants grown from these seeds were placed in light cabinets where the PAR was kept constant and the R : FR was varied from 3.9 (WL) to 0.39 (WL + FR) by the addition of light from tungsten bulbs suitably filtered to allow transmission of FR wavelengths only. The plants from the two seed populations showed different growth strategies although they reached the same overall height in both the WL and WL + FR light regimes. Both populations were able to make a shade-intolerant growth response resulting in a 55 per cent increase in overall height. However, the hedgerow population showed greater extensibility in the younger internodes than did the field population, and the field population showed greater extensibility in the older internodes than did the hedgerow population. It was concluded that these growth strategies would give advantage to the individuals in their respective habitat. The hedgerow population was more suited to competing with an established canopy, whereas the field population was more suited to competing later in the season if the plant became overtopped by surrounding vegetation. Although these two populations are very similar morphologically, they can be separated not only by their shade internode responses but also by the dormancy status and germination requirements of their seed.[106]

The complexity of the natural light environment and its interaction with shade-intolerant species was approached by Morgan and Smith (1978) in an investigation of the effects on stem growth of sunflecks.[107] The characteristics of sunflecks are described elsewhere (p. 18) but these authors point out that the simulated sunflecks used in their experiments were of different quality from those found in the natural environment. The simulated flecks, to which they subjected *Chenopodium album*, contained significantly less B than do natural sunflecks. Nonetheless, switching the plants instantaneously from simulated woodland shade, low irradiance white plus FR (PAR = 25 μmol m^{-2} s^{-1} and R : FR = 0.24) to simulated sunfleck, high irradiance white (PAR = 119 μmol m^{-2} s^{-1} and R : FR = 0.8), caused a 48-per-cent decrease in the stem extension rate, which was measurable after 15 min. The response had a lag phase of about 5 min. When transferred back to shade-simulated light the growth rate increased after a lag of 17 min. If the plants were equilibrated at high fluence and transferred to simulated shade, an increase in growth rate of about 50 per cent was found, with similar lag phases. Changes in growth rate are also found when only PAR is varied. These changes are not as large as those caused by ϕ but must indicate either the involvement of photosynthesis or BAP in this process.

In a study of the stem growth of de-etiolated *Sinapis alba*, the plants were supplied background WL, and FR was added to the stem via a system of glass fibre optics.[108] Increases in stem growth could be detected by LVDT (see p. 72) after 10 to 15 mins. Interestingly, when the leaves were given FR instead of the stem, the lag phase was much greater, viz. 3 to 4 h. Because, in the natural situation, leaves are more likely to be illuminated by short-lived sunflecks than is the stem, the longer lag phase may act as a device which enables the plant to ignore transient signals of this nature. Further, the lag may indicate that there is a hormonal link between the site of photoperception and the site of response, the lag being the time required for transport of such a hormone. It might also be suggested that the internode rather than the leaf would be more likely to detect reflected light of low R : FR from developing nearby vegetation, thus stimulating a rapid, competitive growth response. Ballare *et al.* (1987)[109] have detected redistributions in growth in the seedlings of the shade-intolerant species stimulated by reflected FR-rich radiation from plants growing nearby. They suggest that this might act as an early warning system of imminent competition (p. 82).

This work has been taken a stage further by Child and Smith.[110] These workers showed that if FR was given directly to the internode of plants in background WL, then after a lag phase of about 10 min the extension rate increased, but this increased rate would return to the original rate of 16 min after the FR had been switched off. Casal and Smith (1988)[111] went on to show that if the treatment was continued and the Pfr/Ptot maintained for 45 min, then a long-term growth effect could be triggered. The magnitude of the long-term growth effect was found to be proportional to the Pfr/Ptot and lasted for up to 24 h irrespective of the subsequent changes in the Pfr/Ptot. The site of perception for this long-term effect was found to be the leaf, and the findings are consistent with the view that a growth-promoting substance is released from the leaves as a result of the low Pfr/Ptot and this is transmitted to the stem where it evokes a growth response over an extended period. These workers suggest that the significance of these findings may be related to the non-homogeneous nature of canopy light. This species is able to respond to the predominant environmental signal, rather than to more transient signals, and may maintain high growth rate, reducing the probability of neighbour shade.

LVDT studies of the photocontrol of internode *Vigna sinensis*[112] show that inhibition in WL is fluence-dependent between 0 to 70 Wm^{-2} but that this inhibition is biphasic. First, there is a transient stimulation of growth for 1 h which gives way to the first phase of high inhibition. This itself is replaced by a lower rate of inhibition beginning 2 to 3 h after the onset of light. This pattern of response is mirrored by light of different wavelengths, but it must be mentioned that although WL, R and yellow-green give growth response predominantly inhibitive, B produced a rapid, transient inhibition which gave way to a prolonged promotion of growth rate and returned to the dark control rate only at about 9 h. By giving dichromatic irradiations of B and R these workers are able to closely reproduce the effect of WL, suggesting that there is an interaction between phytochrome and BAP. Experiments in which internodes were covered indicate that in this system the internodes themselves are the sites of photoperception for these growth responses. These results are not easy to interpret in terms of plant-light strategies. It must be remembered that *Vigna sinensis* is a climbing plant and as such may well be subject to different light environment pressures from plants of a caulescent habit. It is disappointing that FR light was not used in these experiments, an omission which detracts from the interpretation.

Effects of reflected radiation

Ballare *et al.* (1987)[109] have been able to detect redistributions in growth in the seedlings of the shade-intolerant species *Chenopodium album*, *Datura ferox* and *Sinapis alba*. These redistributions were found in plants grown in full sunlight but near a stand of grass which supplied reflected FR-rich radiation. These workers found that *S. alba* grown on the sunlit northern side of the grass (these experiments were carried out in the southern hemisphere) developed longer internodes and had a lower leaf: stem dry weight ratio than plants growing in the vicinity of bleached stands of grass. Similar responses were found with the above species when they were grown in full sunlight and supplied with low fluences of far-red reflected by selective mirrors. They suggest that the ability to respond to reflected radiation might act as an early warning system to the plant of forthcoming competition. This is not the only recent report of the effects of reflected radiation. Kasperbauer (1987)[113] has investigated the involvement of the light environment in various effects reported by soybean crop yields. It had previously been reported[114] that when *Glycine max* was grown on well-drained soil, yields were higher from north-south rows when the crop was not water stressed during growth and development as a result of irrigation and higher in east-west rows when the crop was not irrigated and was subject to intermittent water stress. Kasperbauer found that plants grown in rows received light environments which varied in their R : FR depending on the contribution of reflected FR from nearby vegetation. The FR fluence depended upon the nearness and orientation of nearby vegetation and was also shown to be affected by heliotropic movements of leaves at certain row orientations;. These effects became amplified with low solar elevation at the end of day. *Glycine max* was shown to make phytochrome-controlled internode extensions under low R : FR and made fewer side branches. Dry matter accumulation was redistributed as a result of the shade strategies exhibited and this affected, among other parameters, the commercially important dry matter content of the seed.

Tiller formation in grasses

Lolium multiflorum was grown under R : FR typical of the open situation and moderate shade (R : FR = 0.84) while PAR was kept constant.[115] In both situations the ability to produce tillers decreased as the canopy produced by the experimental plants increased. This may have been due to the low R : FR at the base of the plants caused by the canopy, but it was suggested that a limitation in energy supply resulting from the decreased proportion of incident radiation (PIR) available per tiller was in part responsible for the decrease in tiller formation. Furthermore, the number of tillers produced was also reduced at moderate R : FR while the dry weight accumulation by the plant was unaffected. The low R : FR treatment also advanced the reproductive phase and increased the number of fertile tillers per plant. In a later publication, Casal *et al.* show that small reductions in the Pfr/Ptot typical of those which would occur in moderate shade could be shown to have linear relationship with the square root of the tillering rate and in consequence the site-filling rate.[116] They also showed that the plant bases were the site of perception for the tillering response since the irradiation of the plant bases with low-fluence R increased tillering in plants grown under low R : FR to the level found in plants grown under high R : FR. If the bases of the plants grown at high R:FR were treated with low R:FR, then the tillering rate was reduced. They suggested that since the responses were triggered with only a small deviation in the R : FR experienced in the open situation, *Lolium multiflorum* was equipped with a mechanism for detecting competition for light at an early stage.

Petioles

Petiolar responses of shade-intolerant species are regarded as of little importance in species with an upright habit in which the shade-avoidance response is usually effectively expressed by internode elongation. Petiolar extension is, however, a shade-avoidance response in some plants with a rosette habit.[103] Of the plants studied, *Oxalis* was an exception in that it showed a slight decrease in petiole length when grown under a light regime of low R : FR, whereas the other rosette plants, *Tripleurospermum maritimum*, *Medicago arabica*, *Geum urbanum* and *Silene dioica*, all showed an increase in petiole length with increasing simulated shade conditions. To this list might be added *Rumex obtusifolius*,[117] in which petiole length was unaffected by fluence rate but increases were detected when ϕ was lowered. It was also reported that petioles increased in dry weight under low ϕ, indicating a significant change in the use of resources.

Leaves as light traps

Between the atmosphere and the surface of the leaf there is an abrupt change in refractive index (n). The optical boundary thus formed between the air (n = 1) and the leaf tissue (n = 1.5) causes light beams to be reflected or refracted. The structure of leaves of higher plants often shows various degrees of adaptation, exploiting one or other of these effects. Desert plants, for example, have developed pubescent leaf surfaces and as a result of the presence of these fine hairs, leaf reflectance is increased, thereby reducing the level of energy absorbed. In other species the epidermal cells of the upper leaf surface are frequently found to be papillose, that is, the outer walls of the epidermal cells are unusually shaped and protrude from the leaf surface. These protuberances are frequently found in species such as *Anthurium* and *Begonia*, which undergrow tropical canopies. It has been shown that such epidermal cells act as lenses and are able to concentrate diffuse radiation $\times 2$ and collimated light by up to $\times 20$.[118] This focussing may serve to increase the rate of photosynthetic electron transport in the mesophyll cells beneath.

Further boundaries exist within the plant. In contrast to the palisade mesophyll, the spongy mesophyll beneath contains many large air spaces (30 to 40 per cent of leaf volume) which allow the cell walls to act as reflective surfaces. As a result, photons suitable for absorption by the photosynthetic pigments may be intercepted by the chloroplast after reflection back into the palisade layer.[119] Although reflection within the plant is an important phenomenon, Vogelmann and Bjorn suggested[120] that it is the heterogeneity of the plant as an optical medium which leads it to become a light trap. They explain how, without infringing the laws of thermodynamics, it is possible that in the absence of appreciable absorption a plant may contain 3 to 4 times as much light on the inside as is available on the outside. Direct sunlight consists predominantly of collimated light but as it passes into the leaf it becomes progressively scattered. The sub-cellular organelles, cytoplasm, vacuoles, cell walls and intercellular spaces have refractive indices between 1.36 and 1.52 and, in consequence, as light passes from one medium to another it is refracted, diffracted or reflected. As a result of this, when the combined rates of absorption and transmission from the leaf are lower than the rate at which light enters the leaf, then light is collected or retained (see also cotyledons as light traps, p. 75).

Light and lamina development

The prime function of a leaf is to photosynthesise, and since the fluence available for this purpose varies with location, among other factors, it is hardly surprising that many plants have evolved mechanisms whereby fluence can be measured and adaptations made to the development and metabolism of the leaf. The effect of light on a leaf varies with species. At one extreme exists the obligate sun plants, which enjoy an open habitat and are unable to adapt to shade habitats. At the other extreme are the obligate shade plants, which are damaged by exposure to full sunlight. The obligate sun or shade plant develops leaves with little or no adaptability and is consequently limited in habitat. Sun plants are normally capable of higher rates of photosynthesis in full sunlight than are shade plants, but shade plants photosynthesise with greater efficiency at low fluence rates than do sun plants. It is not unusual to find obligate shade plants which photosaturate at fluence rates about 5 per cent of full sunlight. In other plants different degrees of acclimation can be found in leaf development and photosynthetic apparatus.

The strategies of leaf adaptation are often related to the size of the species or individual. All the leaves of a small species or individual are liable to be influenced by a similar light environment and develop very similarly. Bigger plants, trees being the extreme example, support both sun and shade leaves to gain maximum photosynthetic advantage of the light within a self-imposed canopy. Indeed, the high rates of dry matter accumulation found in coniferous trees can be shown to result from photosynthetic adaptation during the development of the needles to the light environment in the particular location within the canopy. In other words, the needle develops either as a sun, an intermediate or a shade needle according to the nature of the light in its immediate vicinity. [121]

The development of the lamina of the leaf can be affected by both the quality and the quantity of light which the leaf receives. Most of the literature concerns the effects of light quantity on leaf morphology, although reports of phytochrome-mediated changes do appear from time to time.

At high fluence rates broad-leafed adaptable plants generally develop smaller, thicker leaves, whereas lower fluence rates result in the development of thinner leaves, e.g. *Fagus sylvatica*, [122] *Phaseolus vulgaris* [123] and *Triticum aestivum*. [124] However, it is not difficult to find exceptions to this rule. *Veronica persica*, for example, [125] shows reduced leaf area as a result of development in neutral density shade and daylight filtered through cinemoid to produce FR, and *Rumex obtusifolius* produces smaller leaves when grown under simulated shade or high and low fluence as produced by fluorescent lights. [117] The suggestion has been made [126] that maximum leaf size might be produced by a particular PAR value for each species and that above and below this value smaller leaves would be produced. Some of the confusion in the literature could be due to the spurious comparison of experiments carried out in the natural environment to those carried out under artificial lighting. The absolute value of natural and artificial fluence rates is of prime importance when comparing results of different authors. Little is yet known of the control mechanisms involved but it might be suggested that below certain PAR values the individual lacks sufficient resources in terms of photosynthate to exploit the larger, thinner leaf strategy.

Sun and shade leaves

Shade leaves

When shade is imposed early in the development of a leaf of a species capable of acclimation, the common response is to produce thinner leaves with larger surface areas (see above). The thinness of such leaves is due to a shortening of the palisade mesophyll cells. Intercellular spaces are large in shade leaves, cell walls thin and epidermal cells long and narrow (Wylie, 1951).[127] Shade causes facultative shade plants to reduce unit leaf area (net photosynthesis per unit area of leaf) but increase in leaf area ratio (total leaf area per total dry mass of plant). Generally, shade plants have less Rubisco in their leaves than do sun plants. This results in lower net photosynthesis and slower growth.

Alocasia macorrhiza is a plant which grows on the floor of Australian tropical rain forests. Although this species contains nearly as much chlorophyll per unit area as a sun plant, only low photosynthetic rates are achieved since photosaturation occurs at low fluence rates.[128] Normally shade plants contain chloroplasts with large numbers of thylakoids in the granal stacks. These stacks are oriented in all directions since efficient absorption of scattered light is of great importance at low fluence rates.

When shade and facultative shade plants are placed under high fluence rates above photosaturation of photosynthesis, bleaching of chlorophylls and inhibition of photosynthesis results. The photoinhibition is largely due to a biochemical change in a protein (HBP-32) involved in electron transport. Light operates via the production of peroxide, which causes the change in HBP-32. Oxygen radicals are formed when electrons cannot be transported and these radicals bleach the chlorophylls.

Sun leaves

When high fluence rates are supplied to the developing leaves of facultative sun plants, they normally become smaller and thicker. The thickness is due to the cells of the palisade mesophyll being longer than would occur at lower fluence rates. At very high fluence rates, cell division is stimulated in the leaves of some plants, causing the development of additional layers of palisade mesophyll. The spongy mesophyll becomes greatly reduced, as do the intercellular spaces.[129,130,131] Turrell (1936)[132] first used the mesophyll area/leaf area ratio (A_{meso}/A) and showed that shade leaves have a low ratio and sun leaves a high ratio. When *Hyptis emoryi* was given different fluence rates the A_{meso}/A was 13 for shade leaves and 40 for sun leaves; these differences were due to cell proliferation in the mesophyll rather than changes in shape.[133] Work with *Fragaria vesca* shows that the anatomical characters of the leaf are determined by total daily fluence.[129] As total daily fluence increases, so does leaf thickness, specific leaf fresh weight, mesophyll cell volume and A_{meso}/A.

Sun leaves exhibit greater rates of transpiration, respiration and photosynthesis than shade leaves[134,135] and are able to dissipate heat more effectively.[136] McCain *et al.* (1988)[137] report that sun leaves of *Acer platanoides* contain 70 per cent more water per unit area than do shade leaves of the same individual. The two types also compartmentalise the water within the leaf in a different fashion. The chloroplasts in the sun leaves contain only 17 per cent of the available water whereas the shade leaf chloroplasts contain as much as 47 per cent of the leaf water. If two equal areas of leaf are compared, then the chloroplasts of the shade leaf contain 60 per cent more water than those of the sun leaf. Sun plants have higher compensation points than shade plants in temperate regions; their photosaturation is only achieved by fluence values close to full sunlight. These higher rates of photosynthesis lead to greater electron transport and sun plants have been shown to have enough carrier capacity to deal with these flow rates.

An analysis of the chemical composition of sun and shade leaves shows that they are in different physiological states.[138] When compared on a per cent dry weight basis, sun leaves of *Fagus* show greater accumulation of lipids, starch, soluble carbohydrate and cutin, as compared to shade leaves, whereas shade leaves contain higher amounts of protein and amino acids.

Phytochrome control of leaf shape in certain dicotyledons

Kasperbauer[139] and Sanchez[140] have worked with young plants of *Nicotiana tabacum* and *Taraxacum officinale*, respectively. By giving the plants end-of-day FR treatments young leaves developed with increased length to breadth ratio. This effect is negated if R is given after the FR.

When *Rumex obtusifolius* was subjected to both variable PAR and R : FR it was found that lowering either of these parameters caused a decrease in the leaf area developed per plant and a similar effect was reported for leaf dry weight.[117] It may be argued that *R. obtusifolius*, *T. officinale* and, to some degree, young *N. tabacum* display a rosette habit and that these responses are restricted to plants of this habit.

Satter and Wetherell[141] found that when *Sinningia speciosa* was given end-of-day FR, it resulted in more erect leaves of decreased area and lower chlorophyll content. In these early days, it was suggested that the lower chlorophyll content was due to a lack of photosynthate, but a more direct effect of phytochrome on chlorophyll synthesis could not be eliminated. Different behaviour was demonstrated in *Taraxacum officinale*.[142] Larger leaves develop when this plant is grown under conditions of low Pfr/Ptot. This being a rosette species, it was suggested that the absence of competition between leaves and stem for photosynthate allowed the development of larger leaves.

Casal *et al.*[143] worked with *Petunia axilaris*, which displays a rosette habit early in development and an upright stem with lateral branches later. End-of-day FR treatment resulted in leaves of greater dry weight and leaf area and lower chlorophyll content. From their results these authors argue that the lower chlorophyll content is more likely to be a direct effect of phytochrome on synthesis than a limitation of resources.

Ecological significance of PAR and phytochrome control

Taylor and Davies (1988)[126] have carried out studies of the effect of varying PAR and R : FR on leaf growth of young plants of the tree species *Betula pendula* and *Acer pseudoplatanus*. They suggest that there is a link between the level of photosaturation of photosynthesis and control of leaf extension and that this may be of ecological significance to these species.

When *Betula*, which had been grown under a fluence of $250 \,\mu\mathrm{mol\,m^{-2}\,s^{-1}}$ on a 16-h light to 8-h dark cycle, was transferred to a variety of fluence rates between 0 to $680 \,\mu\mathrm{mol\,m^{-2}\,s^{-1}}$, changes in growth rate could be detected within 3 h. By measuring the extensibility of cell walls (WEX) using the boiling methanol technique[144] it was possible to show that reducing the fluence rate received by the leaf resulted in reduced leaf extension and a lowering of WEX. These workers[126] suggest that since cell wall extensibility, leaf extension and photosynthesis are all decreased by low PAR, they may be causally related in *Betula* leaves. Increases in WEX are often dependent upon an increased supply of protons to the apoplast.[145] If proton supply limits wall extensibility, then it is possible that this is dependent upon the supply of ATP from photophosphorylation.

This relationship does not appear to operate for *Acer*.[126] The leaves of this species showed little response to changes in PAR either in terms of leaf extension or WEX. *Acer* photosaturates at lower fluence rates than does *Betula*. These characteristics might affect the survival of the seedlings of these species developing under the low PAR levels found in woodland canopies. At low photosynthetic rates the extension of the *Acer* leaf would not be inhibited, and survival would be enhanced, whereas the persistence of *Betula* under these conditions would be adversely affected by the interaction of PAR and leaf extension.

Young plants of both species were also grown at low PAR (25 μmol m^{-2} s^{-1}) and low R : FR (0.26) for 28 days to simulate growth within the canopy. This treatment resulted in a reduction of leaf extension rate and final leaf size in both species. The possibility of BAP involvement in the control of *Betula* leaf extension is not discussed.

Leaves of grasses

When the leaves of grasses are grown under low fluence there is a general tendency for them to develop with increased leaf area and leaf length and decreased leaf width and leaf thickness, as compared with plants grown at high fluence.[124,146] Additionally, it has been shown that the lamina of wheat is reduced in length under high fluence because of the length and number of the bulbiform cells (i.e. epidermal and mesophyll cells).[147] Increase in leaf thickness as a result of high fluence is due to an increase in mesophyll cell number.

The control of these processes is not understood in detail. It has been suggested that the cessation of cell extension in the lamina, after the lamina has emerged from the sheath formed by the older leaves, is controlled by phytochrome.[148,149] Casal *et al.* (1987)[150] showed that when *Lolium multiflorum*, *Sporobolus indicus* and *Paspalum dilatatum* were treated either with end-of-day FR or low R : FR throughout the photoperiod the plants developed longer leaves and, as a result of this, longer shoots. This effect is primarily due to the increase in length of the sheath rather than the blade of the leaf.

Casal and Alvarez (1988)[151] have made an attempt to elucidate the influence of B on *L. multiflorum*. Growing these plants under the high fluence of the natural environment under clear skies at low latitude, they found that if they supplemented natural daylight with various fluence rates of B, the leaf sheath was inhibited in its growth and this inhibition increased with increased fluence of B. The growth of the leaves was unaffected during the early part of development but, after they had attained between 35 and 70 per cent their final length, inhibition of growth became apparent.

Light and chloroplast development

The interaction of light and plastid development is a complex area. Phytochrome control of specific processes has been demonstrated particularly in the synthesis of chlorophyll. On the other hand, other processes, particularly ultrastructural changes, may also be influenced by high fluence and perhaps by BAP. The differences of composition and function between sun and shade chloroplasts are listed in Table 5.1. Sun chloroplasts are found in leaves which have developed under high fluence rates and there is some evidence to indicate that the B component is the salient feature of this radiation, but an unambiguous demonstration of BAP involvement has not yet been made.

Chloroplasts develop from proplastids. These are simple organelles comprising a double membrane, which contains only a few internal membranes or vesicles and are found in young cells in meristematic tissue. In etiolated tissue, proplastids develop into etioplasts. These are approximately twice the size of the proplastid and contain a rudimentary membrane system and one or more prolamellar bodies (paracrystalline lipid required for membrane synthesis). Etioplasts contain a small amount of protochlorophyllide (p-Chlide), Rubisco and related enzymes.[152] When etioplasts are exposed to light, ultrastructural changes take place quite rapidly. The prolamellar body breaks up and vesicles and tubules appear. This event is thought to be closely associated with the photoconversion of the small amount of p-Chlide already present in the developing plastid.[152] After a few hours of high irradiation, more membranes develop and these become organised into the granal stacks of the mature plastid. Chloroplasts have internal membranes containing chlorophyll, and the

Table 5.1   A comparison of the composition and function of sun and shade chloroplasts.

	Sun	Shade
Chloroplast size	Normal	Large
Stromal/thylakoid volume	High	Low
Thylakoid membranes per chloroplast	Low level	High level
No. of thylakoids per grana	Few	Many
Appressed: non-appressed membranes	Low	High
Chlorophyll content per chloroplast	Low	High
Chlorophyll a/b ratio	High	Low
Xanthophylls/β-carotene	Low	High
P680/Chl	High	Low
Cytf/Chl	High	Low
P700/Chl	Little Different	
PSII and PSI activities	High	Low
Fluence for electron transport saturation	High	Low
Quantum yields for whole-chain electron transport	Same	Same
ATP synthetase activity/Chl	High	High

From Anderson *et al.* 1988. *Aust. J. Plant Physiol.* 15 11–26.

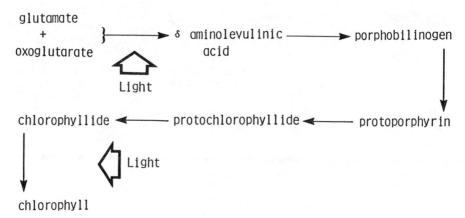

Figure 5.6 Biosynthetic sequence of chlorophyll.

development of etioplast to chloroplast requires light to stimulate the conversion of p-Chlide, by a number of steps, to chlorophyll (see Figure 5.6). In the appropriate situation, e.g. in petals and fruit tissue, the chloroplast may develop into a chromoplast. This is a larger, less structured organelle. Etioplasts in storage tissues may develop into amyloplasts storing starch, or into leucoplasts. These are colourless plastids found in both epidermal and storage cells. All the above structures can be referred to generically as plastids.

Light effects in this area are complex. Evidence can be found in the literature to show that light absorbed by phytochrome, BAP and p-Chlide influences the development of the chloroplast. The effect that light has on this complex system depends on the stage of development of the plastid, light quality and quantity and the tissue under investigation. In roots and underground stems, for example, B light is required to initiate chloroplast development and, although R will enhance the action of B, it will not substitute for it.[153] By contrast, in dark-grown leaves the B-requiring processes which would appear to involve RNA synthesis and metabolism[154] are replaced by a series of non-light-requiring enzymically controlled steps. However, in such tissue a requirement for R remains.[155] R is absorbed first by p-Chlide directly to aid its conversion into chlorophyllide (Chlide) and second by phytochrome, which accelerates chlorophyll synthesis.

If high-irradiance B is available during the conversion of the etioplast into a chloroplast, as it often is in buds developing in the natural environment, sun chloroplasts develop. The differences in sun and shade chloroplasts are summarized in Table 5.1, but it might be mentioned here that sun chloroplasts are smaller than shade chloroplasts and have few grana but high levels of P700, redox coenzymes and Rubisco. The level of the latter is frequently three or four times higher in sun than in shade chloroplasts.[135,156] The effect of B is more apparent in developing plastids in mature tissue than it is during the greening process of etiolated tissue. The formation of shade chloroplasts with larger, more numerous granal stacks appears to be favoured by R and low-light intensities.

Photocontrol of chlorophyll synthesis

The synthesis of chlorophyll and accessory pigments has been extensively studied both in seedlings[157] and in mature plants.[158] Current knowledge of the synthesis and assembly of the components of the photosynthetic apparatus and the control of this process is now very detailed[157,158] and only a summary will be attempted here.

The synthesis of chlorophyll in the majority of angiosperms is light-dependent (see Figure 5.6). Reports of limited chlorophyll synthesis in darkness[159] require the use of pre-irradiated plants. The photoperception for chlorophyll synthesis is via phytochrome and p-Chlide-reductase as shown in Figure 5.6 and discussed below. Only in unusual cases such as tissue cultures and greening roots has B photoreception been shown to be important.[160,161] In the natural environment, this would imply that chlorophyll is not synthesised in the early stages of germination and seedling development if the entire plant is below the soil surface.

Koski *et al.*[162] showed that it was the reductive step in the biosynthetic pathway between p-Chlide and chlorophyllide-a (Chlide-a) which was dependent on the absorption of light by p-Chlide. Much later[163] it was shown that this photoreduction occurred under the influence of the NADP-linked p-Chlide oxidoreductase, which forms a photosensitive complex with p-Chlide and other factors.[164] Although p-Chlide has a central role in the development of the chloroplast, it does not act as a photoreceptor in the same way as phytochrome and BAP. The presence of p-Chlide in the etioplast would appear to inhibit pigment synthesis, apoprotein synthesis and ultrastructural development. In consequence, the level of p-Chlide remains below this inhibitory threshold throughout the greening process.

The role of phytochrome in this system has been demonstrated by a number of workers. Pfr has been shown to influence the size and organisational state of the prolamellar body.[165,166] The galactolipid content of the plastid membrane is also increased by Pfr.[167] The rate at which the Shibata shift occurs is also strongly controlled by phytochrome. The Shibata shift[168] is an *in vivo* spectral shift in the light absorption of chlorophyll-a (Chl-a) that follows p-Chlide conversion to Chlide-a, which is thought to reflect organisational changes with respect to the newly formed chlorophyll and the proteins to which it is attached.

The synthesis of chlorophyll-b (Chl-b) has been shown to be controlled by phytochrome in a number of species,[169] and it is noteworthy that in seedlings of *Chenopodium rubrum* developing under light of different qualities the synthesis of Chl-b requires R, while that of Chl-a is enhanced by B.[170] Such a finding is in keeping with the high Chl-a/b ratio found in sun chloroplasts (see Table 5.1).

The level of the enzyme Rubisco is central to CO_2 fixation in C3 plants and has been shown to be affected by light in a number of ways.[171] Regulation of the activity of this enzyme by light is by no means clear and it has been suggested[172] that some of the conflicting results reported in the literature might be explained if the responsiveness of Rubisco to light varied with the developmental stage of the plant.

Chloroplast membrane organisation

The organisation of the thylakoid membrane is shown in Figure 5.7. The evidence for this model is discussed in the article from which the figure originates.[173] Further information is available in other reviews.[174,175] This model proposed by Andersson and Anderson (1980)[176] resulted from applying the partitioning technique of Albertsson (1971)[177] to the mechanically fragmented granal thylakoid membrane fractions which produced inside-out and right-side-out thylakoid vesicles from appressed and non-appressed membranes, respectively. It had been demonstrated that the appressed membrane fractions were enriched with PSII complex and PSII activity and ATPase is found entirely in the non-appressed regions.[178,179] This led to the novel suggestion of an extreme lateral heterogeneity in the distribution of the photosystems. Previous models had assumed a juxtaposition of the two sets of light-harvesting pigment arrays of PSI and PSII, which would allow the regulation of the excitation energy between the two systems by 'spillover'. This model, however, shows the PSII complex and the associated chlorophyll a/b proteins (LHCII) in the appressed region and therefore segregated from the PSI complex, requiring mobile electron shuttlers between the two systems.

In barley leaves, it has been shown that phytochrome controls the appearance of an m-RNA which codes for the apoprotein of the Chl-a/b protein complex (LHCP).[180] The evidence suggests that this message is produced by a nuclear gene and that a high molecular weight precursor of the apoprotein is produced outside the plastid. The assembly of the LHCP and its incorporation into the thylakoid membranes occur only under continuous illumination. These conditions are necessary to allow the concomitant synthesis of chlorophyll. Phytochrome-mediated regulation of m-RNA translation and *in vitro* transcription of LHCP have been demonstrated in *Lemna gibba*[181,182] and in *Hordeum vulgare*.[180,183]

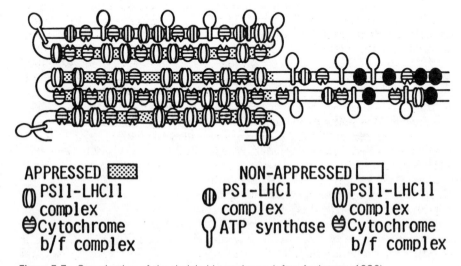

APPRESSED ▨
① PS11-LHC11 complex
❸ Cytochrome b/f complex

NON-APPRESSED ☐
① PS1-LHC1 complex
☿ PS11-LHC11 complex
◊ ATP synthase
❸ Cytochrome b/f complex

Figure 5.7 Organisation of the thylakoid membrane (after Anderson, 1986).

Adaptation of photosynthetic apparatus

The photosynthetic machinery has been found to be adaptive to changes in light quality and quantity in a number of ways and these changes can often be interpreted as forming part of the mechanism whereby plants adapt this area of their activities to the prevailing light environment.

Sun leaves have a higher Chl-a/b ratio, which leads to a higher rate of photosynthesis on a chlorophyll basis and this, together with the fact that sun leaves show photosaturation at higher fluence rates than do shade leaves, indicates that sun and shade leaves have a different photosynthetic apparatus. Sun leaves contain chloroplasts with narrow grana, few thylakoids per granum and fewer thylakoids per stack than do chloroplasts from shade leaves.[122] The very large chloroplasts found in shade leaves are irregularly orientated, may contain as many as 100 thylakoids in a stack and may extend across the entire chloroplast.[184] The ratio of appressed to non-appressed membranes is greater in shade chloroplasts (4 to 5) than in sun chloroplasts (1 to 1.5). Since a leaf transmits only 1 to 5 per cent of incident PAR, the chloroplasts which occur nearer the abaxial rather than the adaxial surface would be exposed to both lower PAR and R : FR. It has been shown that in spinach leaves, in a fixed horizontal position, the ratio of appressed to non-appressed membranes is a linear function of the distance from the adaxial surface.[185] It is suggested that the different pigment ratios displayed by plastids extracted from the different types indicate that the sun chloroplasts possess a smaller antenna and a smaller photosynthetic unit which is adapted for higher rates of quantum conversion than can be achieved by the pigments contained in the chloroplasts of shade leaves.[186] However, in a recent review of this subject it was concluded that the mechanism by which acclimation to low light intensity takes place at the molecular level is still not clear.[174]

Melis and Harvey (1981)[187] and Glick *et al.* (1985)[188] have investigated the development of chloroplasts in a number of both sun- and shade-adapted plants when grown under natural daylight, which were subjected to neutral density and/or interference filtering. They found that growing plants under daylight-deficient conditions or with enhancement of the FR region led to differences in the PSII/PSI reaction centre ratio. FR-enriched plants contained 20 to 25 per cent higher levels of PSII and 20 to 25 per cent lower amounts of PSI reaction centres than plants grown in daylight-deficient light in the FR region. Relative FR enrichment is characteristic of canopy shade light (see p. 16). At first the work of Eskins and Duysens (1984)[189] with soybean seemed to confirm these findings: plants grown under high-irradiance FR had lower levels of pigments and PSI. However, after further examination they found that similar effects could be achieved with low irradiance of R, and they suggest[190] that their results were due to low-level R contamination of their FR source. These workers[190] show that less than 5 per cent Pfr was required to exert phytochrome enhancement of pigment development, membrane polypeptides and ultrastructural aspects of chloroplast development. Once these effects are saturated, the rate of pigment synthesis is of major importance and this is governed by fluence rate. This is also the view of Oelmuller and Mohr (1985).[191] It would appear that the findings of Melis and Harvey (1981) and Glick *et al.* (1985) need to be re-examined.

When plants are grown under B or R, a high Chl-a/b is generally developed under B and a low ratio under R. Similarly, the Chl/P700 is higher in B than in R.[192,193] The fluence response curves for photosynthesis in barley agree with this general hypothesis in that the plants grown in B display sun plant curves while those in R have a shade plant relationship[194]. In a recent review of this subject[195] it was concluded that it is not yet possible to decide whether B or R is responsible for the adaptations of the photosynthetic apparatus which are manifest in sun and shade leaves.

Chloroplast movement

The ability of light to regulate the orientation of the chloroplast in the algae is well known and has been extensively and profitably studied from the point of view of phytochrome action.[196] Evidence from experiments with the filamentous green alga *Mougeotia* indicated that phytochrome is arranged in an organised fashion near the periphery of the cell and probably associated with a membrane such as the plasmalemma.

In higher plants the evidence for chloroplast movement comes from plants of unusual habitat but with characteristics which recommend them as good experimental material. It is obvious that there is a danger of extrapolating from data on B or R stimulation of chloroplast movement in plants such as the floating duckweed *Lemna*, and in particular the aquatic plant *Vallisneria spiralis*, to plants of terrestrial habitat, which enjoy a significantly different light environment (see p. 19).

The orientation of the chloroplast in the aquatic water weeds *Elodea* and *Vallisneria* is known to be controlled by fluence rate.[197,198,199] When plants are kept in darkness, chloroplasts become arranged on all walls of the cell in a random fashion. Light stimulates cyclosis and rotational cytoplasmic streaming, and the chloroplasts move together with the cytoplasm. The final orientation of the chloroplasts depends upon the fluence rate. At low fluence rates chloroplasts become arranged on the periclinal wall, which allows optimal absorption of light under these conditions. At high fluence rates the chloroplasts become arranged on the anticlinal walls, which are at right angles to the incident radiation and, consequently, reduce the level of absorbed energy (Figure 5.8). When chloroplasts move from the high fluence position to

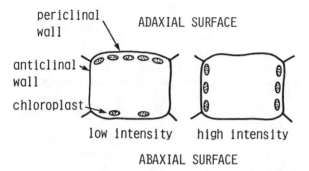

Figure 5.8   Chloroplast arrangement at high and low intensity (redrawn from Seitz, 1987).

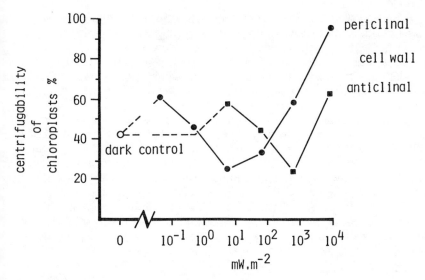

Figure 5.9 Blue fluence response of chloroplast centrifugability at the periclinal and anticlinal walls of *Vallisneria* (after Seitz, 1987).

the low fluence position under low fluence, there is a concomitant increase in the rate of photosynthesis,[200] whereas the movement to the anticlinal walls under high fluence can be predicted to reduce photochemical damage.

Cytoplasmic streaming is thought to be due to a mechanism in the interface between the ectoplasm and the endoplasm. The ectoplasm remains stationary while the endoplasm containing the chloroplasts is moved. The motivation of the endoplasm is thought to be due to the interaction of an endoplasmic myosin with action filaments anchored in the ectoplasm.[201,202] The endoplasmic myosin ATPase interacts with ectoplasmic actin filaments in a cycle of reversible cross bridges, which generates the movement. Ca^{2+} and a supply of ATP have both been shown to be required by this system.[203] The movement of chloroplasts depends on the motive force of the endoplasm and the gliding resistance of, or adherence to, the ectoplasm. The adherence to the ectoplasm, and therefore the ability of chloroplasts to move, can be measured in terms of the centrifugal force required to move the chloroplasts.[204]

The effect of light on the centrifugability of chloroplasts has been investigated in cells of *Vallisneria*.[205] These experiments consider chloroplast from the periclinal wall, where they would congregate at low fluence, and chloroplasts from anticlinal walls, where they would be found under high fluence. The results shown in Figure 5.9 demonstrate that at very low fluence ($<10^{-1} \, mW \, m^{-2}$) periclinal chloroplasts have a 60-per-cent centrifugability, which decreases to 25 per cent as fluence rate increases but increases linearly above $10^{1} \, mW \, m^{-2}$ to 100 per cent at $10^{4} \, mW \, m^{-2}$. Centrifugability shows a similar relationship for chloroplasts from anticlinal walls (where chloroplasts accumulate at high fluence) but there is a phase shift to higher fluence. Minimal centrifugability occurs at $10^{3} \, mW \, m^{-2}$. The difference in these two curves is due to the way light is absorbed as it passes through the cell. The actual fluence in the locality of the chloroplast causes a constant response but the presence of the light absorption gradient determines that at 10 mW m^{-2} the chloroplasts move from the anticlinal wall to the low fluence arrangement on the periclinal wall. At 600 mW m^{-2} the chloroplasts move from the periclinal wall to the high fluence arrangement on the anticlinal wall.

The action spectrum of the high fluence-dependent increase in centri-fugability of chloroplasts in *Vallisneria* has two peaks, at 450 and 370 nm, and a minimum at 400 nm. Seitz[206,207] points to the close correlation between this action spectrum and the absorption spectrum of flavin and suggests that a flavin is the photoreceptor for this system. R (680 nm) has been shown to cause small changes in centrifugability at high fluence. By contrast, the action spectrum for the low fluence change in centrifugability has peaks at 430 and 480 nm and a smaller peak at 680 nm. It has been suggested[204] that this indicates the involvement of Chl-a and Chl-b as photoreceptors. This R-induced activity can be prevented by various inhibitors of photosynthesis. The use of linearly polarised light indicates that the photoreceptors responsible are arranged in an organised fashion probably in association with a membrane.[208]

Leaf orientation

The prime function of the lamina of the leaf is to absorb light for the purpose of photosynthesis. The efficiency of this process is enhanced in many species by the presence of mechanisms which are able to alter the orientation of the leaf. Orientation occurs not only to increase the absorption of light but also to decrease it. In certain species other advantages may result from the ability to alter the position of leaves.

One of the well-known phenomena in this area is the nyctinastic movements in *Albizzia julibrissin*.[209,210] A nastic movement is defined as a movement of an organ which is not determined by the direction of the stimulus. *A. julibrissin* has a multi-pinnate leaf and, at the base or proximal end of each pinnule on both the adaxial and abaxial surface, regions of specialized cells known as pulvinus. The cells in these regions are capable of volume changes which, when the plant is subjected to darkness, are capable of reorienting the pinnule pair from the horizontal plane (the open condition) to a near-vertical plane (the closed condition). It is the pulvinus itself which is the site of photoperception for this response.[211] If in the first 4 h of the photophase the plant is returned to darkness, the leaflets close; this movement has been shown to be controlled by phytochrome, but if the dark treatment is given after this period then the phytochrome control seems to be dominated by an endogenous rhythm which closes the leaflets at the appropriate part of the cycle. These and other workers[212,213,214] went on to show that in the open condition the cell vacuoles of the adaxial pulvinus contained high concentration of potassium and chloride, which were lost very rapidly when the plants were placed in darkness early in the photophase. These osmotica become relocated to the vasculature of the pinnule and the cells of the abaxial pulvinus, which are more turgid in the closed condition than the open condition. At high temperature Pfr enhances the closure in darkness but at low temperature phytochrome control is practically eliminated. The opening response, when plants are transferred from dark to light, is not controlled by phytochrome but is responsive to B, which might indicate at least the involvement of BAP. The advantage of nyctinastic movements to the plants which possess them is not clear. Bunning and Moser (1969)[215] have suggested that the leaves fold to reduce the surface area available for the absorption of moonlight energy. This might have adverse effects on circadian rhythms.

Photonasty is also known in plants and has been extensively studied in *Oxalis oregana* by Bjorkman and Powles.[216,217] This is a shade plant which is able to protect itself from sudden, large increases in fluence rate by reorienting its leaflets from a horizontal position downward to a vertical position. This response detects fluence changes of the order 300 to 400 μmol m^{-2} s^{-1} and has a lag phase of about 10 s. The entire response, folding to a vertical position, takes about 6 min. Unlike the nyctinastic response of *Albizzia*, the leaflet movements of *Oxalis* show no obvious phytochrome involvement, being sensitive only to B between 440 and 490 nm. These authors demonstrate that this response avoids damage to the photosynthetic apparatus of this plant which can be caused by high fluence rates.

It has been suggested that the architecture of the pulvinus may contribute to the conversion of a photonastic response into a phototropic response.[218] Under natural conditions the phototropic responses of photonastic leaves have been shown to be integrated in a number of species, including *Macroptilium atropurpureum*,[219] *Melilotus indicus*[220] and *Lupinus succulentus*,[221] to reorient the upper surface of the lamina towards the sun. The integrated nature of these responses can be illustrated by *Melilotus indicus*, whose leaf orientation is under photonastic control during the day and nyctinastic control at night but which also exhibits solar tracking or diaphototropism.[222] These workers show that the perception of the photonastic stimulus can be detected by both the upper and lower pulvinule, resulting in contraction of the side which is exposed to the light and concomitant expansion of the opposite side.

Different species exhibit different levels of fidelity in sun tracking. *Lupinus*, for example, is capable of tracking to within 15° throughout the daylight hours. The site of photoperception has been shown to be the upper surface of the adaxial pulvinules, which specifically respond to B wavelengths. Each of the leaflets responds independently. The leaflets of *Macroptilium* also reorient individually, but under certain situations an integrated response has been noted and it has been suggested that light absorbed by the the petiole might be piped to leaflets receiving sub-optimal stimulation. The diaphototropic response, as leaf orientation at right angles to the suns rays is known, is of obvious adaptive value. Diaphototropic leaves have been shown to intercept much higher levels of PAR than horizontal leaves,[223] and diaheliotropic leaves have added advantage. The added ability to absorb PAR carries a concomitant increase in heat absorption and this has been found in a number of plants which exhibit paraheliotropism.[224] These plants track the sun with their leaves but the lamina is kept parallel to the sun's beams. While these plants do not reap the entire benefits of sun tracking in terms of PAR absorption, they avoid the concomitant water stress. The authors of this work suggested that certain species are capable of altering the angle of interception of the sun's rays during tracking, according to the water and temperature stress experienced by the leaf.

A remarkable aspect of leaf-sun tracking is the reorientation of the leaves during darkness. Despite the apparent lack of a direction stimulus, the reorientation can be remarkably accurate. Research using *Lavatera* shows that the pre-sunrise positioning of the leaves is dependent upon the position of the previous sunrise.[220]

Stomata

Leaves of higher plants develop cuticles which are impervious to both water and CO_2. Whereas the advantage of controlling water loss is obvious, the exclusion of the raw material of growth is undesirable. In consequence, typical leaves have stomata. In dicotyledonous plants these consist of two kidney-shaped guard cells which have differential thickening of the cell walls. The thickening is arranged in such a way that, as the guard cells increase in turgidity, the outer wall increases in length to a greater degree than the inner wall. Pressure exerted by the end walls of the guard cells on each other has been thought to be involved in the opening mechanism. The guard cells of grasses are typically dumb-bell-shaped and exertion of pressure of one guard cell on the other may be of greater importance in the opening mechanism. When guard cells move apart they leave a pore through which water and CO_2 can move along concentration gradients in opposite directions. Stomatal apertures can be influenced by a number of metabolic and environmental factors. These include relative humidity, intercellular CO_2 concentrations, temperature, hormones and light.[225] Fascinating though this subject is, only the control of stomatal aperture by light will be discussed here.

Pore size

The pore size is determined by the guard cell turgidity, which is regulated by the concentrations of K^+ and Cl^- and malate within the guard cells. Factors able to open stomata are able to increase the rate of K^+ and Cl^- movement into the guard cell from the surrounding cells or increase the rate of synthesis of malate from the hydrolysis of starch grains within the guard cells.

Diurnal stomatal movements

In C3 and C4 plants, stomatal pores are open during the day and closed at night. This enables atmospheric carbon dioxide to diffuse near the mesophyll cells, where it is consumed by the process of photosynthesis. At the same time, the gradient of water out of the plant is at its greatest during the day and, although some functions have been ascribed to this water loss, it is also thought to be deleterious to growth if it proceeds too far. In many plants wilting causes stomatal closure by increasing the endogenous ABA level.[225]

The situation in crassulacean acid metabolism (CAM) plants is entirely different.[226] Many CAM plants are native to arid regions of the world and close their stomata during the day as a water-conservation mechanism. This of course excludes environmental CO_2 as a direct source of carbon for the Calvin cycle. The mechanism that has evolved is that CAM plants open their stomata at night when temperatures are low and allow gaseous interchange when water loss is minimal. During this period, CO_2 is fixed by the enzyme PEP carboxylase, and malate produced as a result of this activity is stored in the vacuoles of the photosynthetic tissue. During the day, when the stomata are closed, the elements of CO_2 are transferred from malate to the Calvin cycle. An interesting exception among CAM is *Isoetes*, which lives in a totally aquatic environment but none the less opens its stomata at night. Presumably the advantage is that there is less competition for dissolved CO_2 at night. A hypothesis which seeks to explain the interaction of light with stomatal movement must be capable of modification to explain stomatal movement in CAM plants.

Wavelength dependence of stomatal opening

Studies using *Vicia faba* showed that K^+ uptake and stomatal aperture had a single optima at 455 nm at low fluence rate.[227] At higher fluence rates another peak could be discerned at 650 nm. These findings agree closely with those of other workers, which show a major peak at 450 nm and a minor one at 660 nm with intact leaves of *Xanthium strumarium*.[228] These results,[227,228] and others[229] with *Commelina communis* are interpreted by Zeiger (1986)[230] as indicating that B and R operate through different parts of the opening mechanism. The small response to R operates an opening mechanism by stimulating guard cell chloroplasts, whereas the B response, being much greater, occurs not only via the guard cell chloroplast but also stimulates the B photosystem. Further support for this interpretation comes from experimentation with the stomatal responses of the orchid, *Paphiopedilum*.[231] These guard cells are unusual in that they lack chloroplasts. They respond predominantly to B, and the very limited response to R can be explained in terms of the reduced intercellular CO_2 caused by the photosynthetic activity of the mesophyll. On the other hand, Roth-Bejerano and Nejidat (1987),[232] working with various K^+-channel blockers on the stomata of *Commelina communis*, found that opening and closing could be inhibited in R and that this effect was prevented by supplementing the light with FR, indicating a direct role for phytochrome in K^+ transport.

Guard cell chloroplasts

The stomata which occur in the albino portion of the variegated leaf of *Chlorophytum comosum* have been found to possess guard cells which contain a normal chloroplast content.[233] It was possible to show that the guard cell chloroplasts had the light-harvesting pigments of PSI and PSII and the ability to conduct electron transport and synthesise ATP. This species exhibits a low level of opening under the influence of R, which can be inhibited in epidermal peels by the addition of DCMU, an inhibitor of photosynthetic electron transport, while this opening was insensitive to KCN.[229] It was concluded from this work that the energy required to open stomata was supplied via a photosynthetic process proceeding in the guard cell of the chloroplast. It would be wrong to give the impression that processes in the guard cell chloroplast are fully understood. Evidence to date suggests that the Calvin cycle does not function in the guard cell chloroplast. Furthermore, whether ATP is produced via cyclic or non-cyclic electron transport is not a matter of general agreement.

Photoperception by stomata

When illuminated with B light in the presence of K^+, guard cell protoplasts swell. They will then shrink when placed in the dark or subjected to other closure signals such as ABA treatment.[234] These findings indicate that the guard cells perceive light and move independently of signals from the mesophyll or any other cell. In intact leaves, similar evidence comes from the work of Sharkey and Raschke (1981),[228] who used cyanazine to inhibit mesophyll photosynthesis of *Xanthium strumarium* and showed that light would open stomata in the absence of CO_2 fixation. The opposite approach was used by Zeiger and Field (1981),[235] who saturated the light requirement for photosynthesis with an R laser and obtained an increase in stomatal conductance by adding a low-fluence B laser. The conclusion from this type of work is that the observed relationship between photosynthesis and stomatal aperture is a result of simultaneous response of the mesophyll and the guard cell to light rather than the transference of a message of some description between the mesophyll and the guard cell.

The B responses of guard cells are regarded as operating via the BAP, as is found in other plant responses. It may be surmised that in the natural environment the B response is more important than the R response, not only because of the difference in magnitude of the two responses but also as a result of the kinetics of the response. When the attached leaves of *Malva paviflora* are irradiated with B the increase in leaf conductance is not only much larger but is also more rapid than when they are exposed to R.[236] The stomatal opening of *Paphiopedilum* showed similar insensitivity to R while opening under B.[231] Using measurements of *Malva paviflora* stomata taken in the field, it was possible to demonstrate a pre-dawn opening of stomata. This was correlated with the pre-dawn increases in B reported by Goldberg and Klein (1977),[237] and it was suggested that pre-dawn opening could be of advantage to a plant by enhancing the early morning rates of photosynthesis. This would allow particular advantage to a plant growing in a hot environment where midday closure of stomata due to wilting could markedly reduce daily assimilation rates.

BAP a proton pump?

Zeiger *et al.*(1978)[238] and Zeiger (1986)[230] propose a chemiosmotic mechanism for stomatal movement. They suggest that ion fluxes in guard cells depend on proton extrusion and this is the single energy-requiring step of guard cell movement. The extrusion of protons from the cell generates a pH and electrical gradient which allows K^+ to enter the cell. Cl^- is taken up in a Cl^-/OH^- antiport. Proton extrusion is postulated to be primarily dependent on the activity of a plasmalemma-bound ATPase. Light enables the guard cell chloroplasts to provide the requisite ATP for proton extrusion during guard cell opening. B causes additional extrusion of protons when the activity of the guard cell chloroplasts is saturated. Whether B stimulates the same ATPase at the plasmalemma or whether there are B-specific ATPases is not known.

Phytochrome and the guard cell

Phytochrome has been shown by immunological studies to occur in guard cells. Holmes and Klein (1986)[239] have shown that phytochrome can influence the endogenous rhythm, which can be demonstrated for the stomatal movement of *Phaseolus vulgaris* when plants are transferred from light/dark transitions to continuous darkness. These workers found that the disappearance of Pfr affected the amplitude of the response in darkness by causing damping out. They found no evidence that phytochrome could influence the phase of the stomatal rhythm.

Leaf abscission

At some stage in their development deciduous leaves develop an abscission zone at the proximal end of the petiole, i.e. where it joins the stem. The abscission zone consists of one or more layers of thin-walled parenchymatous cells which develop from anticlinal divisions across the petiole. From the results of some experiments it is possible to argue that a senescence factor moves from the blade of the senescing leaf to influence the changes which take place in the abscission zone.[240,241] However, a more complicated hypothesis may be required since examples are known where senescence occurs in the autumn and abscission occurs in the spring.[242] This is to say that these processes are usually coupled but can occur independently. Irrespective of the immediate stimulus, the cells of the distal region of the abscission zone secrete pectinases and cellulases from the cytoplasm through the cell wall. These enzymes digest the middle lamella and often the cell walls in this region. Both ethylene production and respiration increase in the abscission zone, and this is thought to be associated with the elongation of cells in the proximal region.

It has been known for many years that light is involved in leaf abscission, mainly as a result of the observation that street lighting in towns retards leaf abscission.[243] More recent work[244,245,246] has demonstrated that dark-induced leaf abscission in cuttings of *Phaseolus vulgaris* and *P. aureus* can be completely inhibited by WL and there is a phytochrome involvement in this effect. If plants were treated with FR prior to the dark treatment, abscission was enhanced. If they were given R prior to darkness, the abscission process was greatly reduced.

Decoteau and Craker[247,248] have shown that the leaves of cuttings of mung bean, *Vigna radiata*, are induced to abscise by dark treatment and are inhibited from doing so by low levels of R. Maximal inhibition is achieved by a 12-h R treatment in each 24-h cycle. It is suggested that the important factors are the photoequilibrium of phytochrome and the length of the dark period. The effect of R is reversed by a few minutes of FR prior to the dark period. In later papers these workers[249,250] show similar effects of R inhibition and FR promotion of dark-induced abscission with cuttings of *Coleus blumei* and suggest a link between R and FR effects on abscission and ethylene production. These findings are in agreement with the hypothesis that photoperception takes place in the blade of the leaf and an inhibitory substance is translocated to the abscission zone.

References for Chapter 5

1   Goeschl, J.D., Pratt, H.K. and Bonner, B.A. 1967. An effect of light on the production of ethylene and the growth of the plumular portion of etiolated pea seedlings. *Plant Physiol.*, **42**, 1077–1080.

2   Kang, B.G. and Ray, P.M. 1969. Ethylene and carbon dioxide as mediators in the response of the bean hypocotyl hook to light and auxins. *Planta.*, **87**, 206–216.

3   Bewley, J.D. and Black, M. 1982. *Physiology and biochemistry of seeds.* Vol. I. Berlin, Springer-Verlag.

4   Mandoli, D.F. and Briggs, W.R. 1982a. Optical properties of etiolated plant tissues. *Proc. Nat. Acad. Sci. USA*, **79**, 2902–2906.

5   Mandoli, D.F. and Briggs, W.R. 1983. the physiology and optics of plant tissues. *Whats New in Plant Physiol.*, **14**, 13–16.

6   Vogelmann, T.C. and Haupt, W. 1985. The blue light gradient in unilaterally illuminated irradiated maize coleoptiles: measurement with fibre optic probe. *Physiol. Plant*, **41**, 569–576.

7   MacLeod, K., Digby, J. and Fern, R.D. 1985. Evidence inconsistent with the Blaauw model of phototropism. *J. Exp. Bot.*, **36**, 312–319.

8   Thomson, B.F. 1950. The effect of light on the rate of development of *Avena* seedlings. *Am. J. Bot.*, **37**, 284–291.

9   Blaauw, O.H., Blaauw-Jensen, G. and Van Leeuwen, W.J. 1968. An irreversible red-light induced growth response in *Avena. Planta.*, **82**, 87–104.

10   Warner, T.J. and Ross, J.D. 1981. Phytochrome control of maize coleoptile section elongation: The role of cell wall extensibility. *Plant Physiol.*, **68**, 1024–1026.

11   Shinkle, J.R. 1986. Photobiology of phytochrome-mediated growth responses in sections of stem tissue from etiolated oats and corn. *Plant Physiol.*, **81**, 533–537.

12   Fuhr, G., Bleiss, W. and Goring, H. 1980. Phytochrome-regulated growth of excised coleoptile tips of *Triticum aestivum* induced by blue, red and far-red light. *Plant Cell Physiol.*, **21**, 571–580.

13   Lawson, V.R. and Weintraub, R.L. 1975a. Effects of red light on the growth of intact wheat and barley coleoptiles. *Plant Physiol.*, **56**, 44–50.

14   Roesel, H.A. and Haber, H.A. 1963. Studies of effects of light and of gibberellin sensitivity in relation to age growth and illumination in intact wheat coleoptiles. *Plant Physiol.*, **38**, 523–532.

15   Bleiss, W. and Smith, H. 1985. Rapid suppression of extension growth in dark grown wheat seedlings by red light. *Plant Physiol.*, **77**, 552–555.

16   Lawson, V.R. and Weintraub, R.L. 1975b. Interactions of microtubule disorganisers, plant hormones and red light in wheat coleoptile segment growth. *Plant Physiol.*, **55**, 1062–1066.

17   Pjion, C.T. and Furuya, M. 1967. Phytochrome action in *Oryza sativa*. I. Growth responses of etiolated coleoptiles to red, far-red and blue light. *Plant Cell Physiol.*, **8**, 709–718.

18   Pjon, C-J, and Furuya, M. 1968. Phytochrome action in *Oryza sativa*. II. The spectrophotometric versus the physiological status of phytochrome in coleoptiles. *Planta.*, **81**, 303–313.

19   Furuya, M., Pjon, C-J., Fujii, T. and Ito, M. 1969. Phytochrome action in *Oryza sativa*. III. The separation of photoreceptive site and growing zone in coleoptiles and auxin transport as effector system. *Dev. Growth Diff.*, **14**, 95–105.

20   Blaauw, O.H., Blaauw-Jensen, G. and Van Leeuwen, W.J. 1968. An irreversible red-light induced growth response in *Avena. Planta.*, **82**, 87–104.

21   Vanderhoef, L.N., Quail, P.H. and Briggs, W.R. 1979. Red light-inhibited

mesocotyl elongation in maize seedlings. II. Kinetic and spectral studies. *Plant Physiol.*, **63**, 1062–1067.

22 Mandoli, D.F. and Briggs, W.R. 1981. Phytochrome control of two low irradiance responses in etiolated oat seedlings. *Plant Physiol.*, **67**, 733–739.

23 Mandoli, D.F. and Briggs, W.R. 1982b. Photoperceptive sites and the function of tissue light-piping in photomorphogenesis of etiolated oat seedlings. *Plant Cell Environ.*, **5**, 137–145.

24 Shinkle, J.R. and Briggs, W.R. 1985. Physiological mechanism of the auxin-induced increase in light sensitivity of phytochrome-mediated growth responses in *Avena* coleoptiles sections. *Plant Physiol.*, **79**, 349–356.

25 Vanderhoef, L.N. and Briggs, W.R. 1978. Red light inhibited mesocotyl elongation in maize seedlings. I. The auxin hypothesis. *Plant Physiol.*, **61**, 534–537.

26 Van Overbeek, J. 1936. Growth hormone and mesocotyl growth. *Rec. Trav. Bot. Neerl.*, **33**, 333–340.

27 Iino, M. 1982a. Action of red light on indole-acetic acid status and growth of coleoptiles of etiolated maize seedlings. *Planta.*, **156**, 21–32.

28 Iino, M. 1982b. Inhibitory action of red light on the growth of coleoptiles of maize mesocotyls: evaluation of the auxin hypothesis. *Planta.*, **156**, 388–395.

29 Schopfer, P., Fidelak, K.H. and Schafer, E. 1982. Phytochrome-controlled extension growth of *Avena sativa* seedlings. I. Kinetic characterisation of mesocotyl, coleoptile and leaf responses. *Planta.*, **154**, 224–230.

30 Yahalom, A., Epel, B.L. and Glinka, Z. 1988. Photomodulation of mesocotyl elongation in maize seedlings: Is there a correlative relationship between phytochrome, auxin and cell wall extensibility. *Plant Physiol.*, **72**, 428–433.

31 Gaba, V., Black, M. and Attridge, T.H. 1984. Photocontrol of hypocotyl elongation in de-etiolated *Cucumis sativus*. Long term fluence rate dependent responses to blue light. *Plant Physiol.*, **74**, 897–900.

32 Drumm-Herrel, H. and Mohr, H. 1985. Relative importance of blue light and light absorbed by phytochrome in growth of mustard (*Sinapis alba*). *Photochem. Photobiol.*, **42**, 735–739.

33 Elliot, W.M. and Shen-Miller, J. 1976. Similarity of dose responses, action spectra, and red light responses between phototropism and photoinhibition of growth. *Photochem. Photobiol.*, **23**, 195–199.

34 Cosgrove, D.J. 1982. Rapid inhibition of hypocotyl growth by blue light in *Sinapis alba*. *Plant Sci. Lett.*, **25**, 305–312.

35 Gorton, H.L. and Briggs, W.R. 1980. Phytochrome responses to end-of-day irradiations in light grown corn in the presence and absence of Sandoz 9789. *Plant Physiol.*, **66**, 1024–1030.

36 Went, F.W. 1928. Wuchstoff und Wachstum. *Rec. Trav. Bot. Neerl.*, **25**, 1–116.

37 Parsons, A., Firn, R.D. and Digby, J. 1988. The role of the coleoptile apex in controlling organ elongation. I. Effects of decapitation and apical excisions. *J. Exp. Bot.*, **39**, 1331–1341.

38 Parsons, A., Firn, R.D. and Digby, J. 1988. The role of the coleoptile apex in controlling organ elongation. II. Effects of auxin substitution and auxin transport inhibitors on decapitated coleoptiles. *J. Exp. Bot.*, **39**, 1343–1354.

39 Imasekei, H., Pjon, C.J. and Furuya, M. 1971. Phytochrome action in *Oryza sativa*. IV. Red and far-red reversible effect on the production of ethylene. *Plant Physiol.*, **48**, 241–244.

40 Samimy, C. 1978. Effect of light on ethylene production and hypocotyl growth of soybean seedlings. *Plant Physiol.*, **61**, 772–774.

41 Buhler, B., Drumm, H. and Mohr, H. 1978. Investigations on the role of ethylene in phytochrome-mediated photomorphogenesis. I. Anthocyanin synthesis. *Planta.*, **142**, 109–177.

42 Dei, M. 1981. Evidence that ethylene is not involved in red-light stimulation of

chlorophyll formation in etiolated cucumber seedlings. *Plant Cell Physiol.*, **22**, 699–707.

43   Rohwer, F. and Schierle, J. 1982. Effect of light on ethylene production: red light enhancement of 1-aminocyclopropane-1-carboxylic acid concentration in etiolated pea shoots. *Z. Planzenphysiol.*, **107**, 295–300.

44   Janes, H.W., Loercher, L. and Frenkel, C. 1976. Effects of red light and ethylene on the growth of etiolated lettuce seedlings. *Plant Physiol.*, **57**, 440–443.

45   Vangronsveld, J., Clijisters, H. and Van Poucke, M. 1988. Phytochrome-controlled ethylene biosynthesis of intact etiolated bean seedlings. *Planta.*, **174**, 19–24.

46   Meijer, G. 1968. Rapid growth inhibition of gherkin hypocotyls in blue light. *Acta Bot. Neerl.*, **17**, 9–14.

47   Cosgrove, D. 1987. Wall relaxation and the driving forces for cell expansion. *Plant Physiol.*, **84**, 561–564.

48   Virgin, H.I. 1962. Light-induced unfolding in grass leaf. *Plant Physiol.*, **15**, 380–389.

49   Beevers, L., Loveys, B., Pearson, J.A. and Wareing, P.F. 1970. Phytochrome and hormonal control of expansion and greening of etiolated wheat leaves. *Planta.*, **90**, 286–294.

50   Poulson, R. and Beevers, L. 1969. The influence of growth regulators on the unrolling of barley leaf sections. *Plant Physiol.*, **44**, Suppl. 29 137a.

51   Hepler, P.K. and Wayne, R.O. 1985. Calcium and plant development. *Ann. Rev. Plant Physiol.*, **36**, 397–569.

52   Viner, N., Whitelam, G. and Smith, H. 1988. Calcium and phytochrome control of leaf unrolling in dark-grown barley. *Planta.*, **175**, 209–213.

53   Meijer, G. 1958. Influence of light on the elongation of gherkin seedlings. *Acta Bot. Neerl.*, **7**, 614–620.

54   Black, M. and Shuttleworth, J.E. 1974. The role of the cotyledons in the photocontrol of hypocotyl extension in *Cucumis sativus*. *Planta.*, **117**, 57–66.

55   Gaba, V. and Black, M. 1979. Two separate photoreceptors control hypocotyl growth in green seedlings. *Nature*, **278**, 51–54.

56   Attridge, T.H., Black, M. and Gaba, V. 1984. Photocontrol of hypocotyl elongation in light grown *Cucumis sativus*. A synergism between blue-light photoreceptor and phytochrome. *Planta.*, **162**, 422–426.

57   Holmes, M.G. and Wagner, E. 1980. A re-evaluation of phytochrome involvement in time measurement in plants. *J. Theor. Biol.*, **83**, 255–265.

58   Thomas, B. and Dickinson, H.G. 1979. Evidence for two photoreceptors controlling growth in de-etiolated seedlings. *Planta.*, **146**, 545–560.

59   Thomas, B., Tull, S.E. and Warner, T.J. 1980. Light dependent gibberellin responses in hypocotyls of *Lactuca sativa*. *Plant Sci. Lett.*, **19**, 355–362.

60   Wildermann, A., Drumm, H., Schafer, E. and Mohr, H. 1978a. Control by light of hypocotyl growth in de-etiolated mustard seedling. I. Phytochrome as the only photoreceptor pigment. *Planta.*, **141**, 211–216.

61   Kristie, D.N. and Jolliffe, P.A. 1987. A rapid phytochrome-mediated growth response in etiolated *Sinapis alba* hypocotyls. *Can. J. Bot.*, **65**, 2017–2023.

62   Cosgrove, D.J. 1981. Rapid suppression of growth by blue light. *Plant Physiol.*, **67**, 584–590.

63   Cosgrove, D.J. and Green, P.B. 1981. Rapid suppression of growth by blue light. Biophysical mechanism of action. *Plant Physiol.*, **68**, 1447–1453.

64   Moore, K. and Lovell, P. 1970. Differential effects of the embryonic axis on chlorophyll production and photosynthesis of mustard cotyledons. *Planta.*, **93**, 289–294.

65   Oelze-Karow, H. and Mohr, H. 1978a. Die Bedeutungdes phytochroms fur die

Entwicklung der kapazitat fer Photophosporylierung. *Ber. Dtsch. Bot. Gaz.*, **91**, 603–610.

66 Oelze-Karow, H. and Mohr, H. 1986. Appearance of photophosphorytation capacity, threshold versus graded control of phytochrome. *Photochem. Photobiol.*, **44**, 221–230.

67 Oelze-Karow, H. and Mohr, H. 1988. Rapid transmission of a phytochrome signal from the hypocotyl hook to cotyledons in mustard (*Sinapis alba*). *Photochem. Photobiol.*, **47**, 447–450.

68 Hardy, S.J., Castelfranco, P.A. and Rebieiz, C.A. 1971. Effect of hypocotyl hook on chlorophyll accumulation in excised cotyledons of *Cucumis sativus*. *Plant Physiol.*, **47**, 705–708.

69 Sangeetha, B. and Sharma, R. 1988. Phytochrome-regulated expansion of mustard (*Sinapis alba*) cotyledons. *J. Exp. Bot.*, **39**, 1355–1366.

70 Knapp, A.K., Vogelmann, T.C., McClean, T.M. and Smith, W.K. 1988. Light and chlorophyll gradients within *Cucurbita* cotyledons. *Plant Cell Environ.*, **11**, 257–263.

71 Wilkins, M.B. 1984. Gravitropism. In: Wilkins, M.B. (ed.), *Advanced Plant Physiology*. London, Pitman Pub. Ltd, 163–185.

72 Griffiths, H.J. and Audus, L.J. 1964. Organelle distribution in the statocyte cells of the root tip of *Vicia faba* in relation to geotropic stimulation. *New Phytol.*, **63**, 319–333.

73 Suzuki, T., Kondo, N. and Fujii, T. 1979. Distribution of growth regulators in relation to the light-induced geotropic responsiveness of *Zea* roots. *Planta.*, **145**, 323–329.

74 Mertens, M. and Weiler, E.W. 1983. Kinetic studies of the redistribution of endogenous growth substances in gravi-reacting plant organs. *Planta.*, **158**, 339–348.

75 Briggs, W.R. and Siegelmann, H.W. 1965. Distribution of phytochrome in etiolated seedlings. *Plant Physiol.*, **40**, 934–941.

76 Pratt, L.H. and Coleman, R.L. 1974. Phytochrome distribution in etiolated grass seedlings as assayed by a direct antibody-labelling method. *Amer. J. Bot.*, **61**, 195–202.

77 Tepfer, D.A. and Bonnett, H.T. 1972. The role of phytochrome in geotropic behavior of roots of *Convolvulus arvensis*. *Planta.*, **106**, 311–324.

78 Klemmer, R. and Schneider, H.A.W. 1979. On a blue light effect and phytochrome in the stimulation of georesponsiveness of maize roots. *Z. Pflanzenphysiol.*, **95**, 189–197.

79 Mirza, J.I. 1987. The effects of light and gravity on the horizontal curvature of roots of gravitropic and agravitropic *Arabidopsis thaliana*. *Plant Physiol.*, **83**, 118–120.

80 Torrey, J.G. 1952. Effects of light on elongation and branching in pea roots. *Plant Physiol.*, **27**, 591–602.

81 Wilkins, M.B. and Wain, R.L. 1974. The root cap control of root elongation in *Zea mays* seedlings exposed to white light. *Planta.*, **121**, 1–8.

82 Pilet, P.E. and Rivier, L. 1980. Light and dark georeaction of maize roots: effect and endogenous level of abscisic acid. *Plant Sci. Lett.*, **18**, 201–206.

83 Feldman, L.J. 1984. Regulation of root development. *Ann. Rev. Plant Physiol.*, **35**, 223–242.

84 Pickard, B.G. 1985. Roles of hormones, protons and calcium in geotropism. In: Pharis, R.P. and Reid, D.M. (eds.), *Encyc. Plant Physiol. New Series*. Berlin, Springer-Verlag vol 11, 193–281.

85 Masuda, Y. 1962. Effect of light on a growth inhibitor in wheat roots. *Physiol. Plant*, **15**, 780–790.

86   Gaspar, T. 1973. Inhibition of root growth as a result of methyleneoxindole formation. *Plant Sci. Lett.*, **1**, 115–118.

87   Feldman, L.J., Arroyave, N.A. and Sun, P.S. 1985. Abscisic acid, xanthoxin and violaxanthin in the caps of gravistimulated maize roots. *Planta.*, **166**, 483–489.

88   Robert, M.L., Taylor, H.F. and Wain, R.L. 1975. Ethylene production from cress roots and excised cress root segments and its inhibition by 3, 5-iiodo-4-hydroxybenzoic acid. *Planta.*, **126**, 273–284.

89   Chadwick, A.U. and Burg, S.P. 1967. An explanation of the inhibition of root growth caused by indole-3-acetic acid. *Plant Physiol.*, **42**, 192–200.

90   Bucher, D. and Pilet, P.E. 1981. Ethylene production in growing and gravireacting maize and pea root segments. *Plant Sci. Lett.*, **22**, 7–11.

91   Eliasson, L. and Bollmark, M. 1988. Ethylene as a possible mediator of light induced inhibition of root growth. *Physiol. Plant*, **72**, 605–609.

92   Jarvis, B.C. and Shaheed, A.I. 1987. Adventitous root formation in relation to irradiance and auxin supply. *Biol. Plant*, **29**, 321–323.

93   Galston, A.W., Tuttle, A.A. and Penny, P.J. 1964. A kinetic study of growth movements and photomorphogenesis in etiolated pea seedlings. *Am. J. Bot.*, **51**, 853–858.

94   Rubenstein, B. 1971a. The role of various regions of the bean hypocotyl on red light-induced hook opening. *Plant Physiol.*, **48**, 187–192.

95   Rubenstein, B. 1971b. Auxin and red light in the control of hypocotyl hook opening in beans. *Plant Physiol.*, **48**, 187–192.

96   Bjorn, L.O. and Virgin, H.I. 1958. The influence of red light on the growth of pea seedlings. An attempt to localize perception. *Physiol. Plant*, **11**, 363–373.

97   Holmes, M.G. and Smith, H. 1977a. The function of phytochrome in the natural environment. I. Characterisation of daylight for studies in photomorphogenesis and photoperiodism. *Photochem. Photobiol.*, **25**, 533–538.

98   Holmes, M.G. and Smith, H. 1977b. The function of phytochrome in the natural environment. II. III Measurement and calculation of plytochrome photoequilibrium. *Photochem. Photobiol.*, **25**, 539–545.

99   Holmes, M.G. and Smith, H. 1977c. The function of phytochrome in the natural environment. IV. Light quality and plant development. *Photochem. Photobiol.*, **25**, 551–557.

100   Smith, H. and Holmes, M.G. 1977. The function of phytochrome in the natural environment. II. III Measurement and calculation of plytochrome equilibrium. *Photochem. Photobiol.*, **25**, 547–550.

101   Morgan, D.C. and Smith, H. 1976. Linear relationship between phytochrome photoequilibrium and growth in plants under simulated natural radiation. *Nature*, **262**, 210–211.

102   Hartmann, K.M. 1966. A general hypothesis to interpret high energy phenomena of photomorphogenesis on the basis of phytochrome. *Photochem. Photobiol.*, **5**, 349–366.

103   Morgan, D.C. and Smith, H. 1979. A systematic relationship between phytochrome controlled development and species habitat for plants grown in natural radiation. *Planta.*, **145**, 253–258.

104   Morgan, D.C. and Smith, H. 1978. The relationship between phytochrome photoequilibrium and development in light grown *Chenopodium album*. *Planta.*, **142**, 187–193.

105   Bain, A.B. and Attridge, T.H. 1988. Shade-light mediated responses in field and hedgerow populations of *Galium aparine*. *J. Exp. Bot.*, **39**, 1759–1764.

106   Froud-Williams, R.J. 1985. The biology of cleavers (*Galium aparine*). *Aspects of Applied biology (The biology and control of weeds in cereals)*, **9**, 189–195.

107   Morgan, D.C. and Smith, H. 1978. Simulated sunflecks have large rapid effects on plant stem extension. *Nature*, **273**, 534–536.

108 Morgan, D.C., O'Brien, and Smith, H. 1980. Rapid photomodulation of stem extension in light grown *Sinapis alba*. Studies of perception and photoreceptor. *Planta.*, **150**, 95–101.

109 Ballare, C.L., Sanchez, R.A., Scopel, R.A., Casal, J.J. and Ghersa, C.M. 1987. Early detection of neighbouring plants by phytochrome perception of changes in reflected sunlight. *Plant Cell and Environ.*, **10**, 551–557.

110 Child, R. and Smith, H. 1987. Phytochrome action in light grown mustard: Kinetics, fluence rate compensation and ecological significance. *Planta.*, **172**, 219–229.

111 Casal, J.J. and Smith, H. 1988. Persistent effects of changes in phytochrome status on internode growth in light-grown mustard: Occurrence, kinetics and locus of perception. *Planta.*, **175**, 214–220.

112 Lecharny, A. and Jaques, R. 1980. Light inhibition of internode elongation in green plants. A kinetic study with *Vignia sinensis. Planta.*, **149**, 384–388.

113 Kasperbauer, M. 1987. Far-red light reflection from green leaves and effects on phytochrome-mediated assimilate partitioning under field conditions. *Plant Physiol.*, **85**, 350–354.

114 Hunt, P.G., Sojka, R.E., Matheny, T.A. and Wollum, I.I. 1985. Soybean responses to *Rhizobium japonicum* strain, row orientation and irrigation. *Agron. J.*, **77**, 720–727.

115 Casal, J.J., Deregibus, V.A. and Sanchez, R.A. 1985. Variations in tiller dynamics and morphology in *Lolium multiflorum* vegetative and reproductive plants as affected by differences in red/far-red irradiations. *Ann. of Bot.*, **56**, 553–559.

116 Casal, J.J., Sanchez, R.A. and Deregibus, V.A. 1987b. Tillering responses of *Lolium multiflorum* plants to changes of red/far-red ratio typical of sparse canopies. *J. Exp. Bot.*, **38**, 1432–1439.

117 McLaren, J.S. and Smith, H. 1978. Phytochrome control of the growth and development of *Rumex obtusifolius* under simulated canopy light environment. *Plant Cell Environ.*, **1**, 61–67.

118 Bone, R.A., Lee, D. and Norman, J.M. 1985. Epidermal cells functioning as lenses in leaves of tropical rain-forest shade plants. *Appl. Opt.*, **24**, 1408–1412.

119 Terashima, I. and Saeki, T. 1983. Light environment within a leaf. I. Optical properties of paradermal sections of camelia leaves with special reference to differences in the optical properties of palisade and spongy tissue. *Plant Cell Physiol.*, **24**, 1493–1501.

120 Vogelmann, T.C. and Bjorn, L.O. 1986. Plants as light traps. *Physiol Plant*, **68**, 704–708.

121 Jarvis, P.F. 1981. Production efficiency of coniferous forest in the U.K. In: Johnson, C.B. (ed.), *Physiological Processes Limiting Plant Productivity*. London, Butterworths, 81–108.

122 Lichtenthaler, H.K. 1985. Differences in morphology and chemical composition of leaves grown in different light intensities and qualities. In: Baker, N.R., Davies, W.J. and Ong, C.K. (eds.), *Control of Leaf Growth*. S.E.B. Seminar Series. Cambridge, Cambridge University Press, 201–221.

123 Dale, J.E. 1965. Leaf growth in *Phaseolus vulgaris*. II. Temperature effects and the light factor. *Ann. Bot. NS.*, **29**, 293–308.

124 Friend, D.J.C., Helson, V.A. and Fischer, J.E. 1962. Leaf growth in Marquis wheat as regulated by temperature, light and daylength. *Can. J. Bot.*, **40**, 1299–1311.

125 Fitter, A.H. and Ashmore, C.J. 1974. Responses of two *Veronica* species to shaded woodland light climate. *New Phytol.*, **73**, 997–1001.

126 Taylor, G. and Davies, W.J. 1988. The influence of photosynthetically active

radiation and simulated shadelight on the leaf growth of *Betula* and *Acer*. *New Phytol.*, **108**, 393–398.

127   Wylie, R.B. 1951. The principles of foliar organisation shown by sun-shade leaves from ten different species of deciduous dicotyledonous trees. *Am. J. Bot.*, **38**, 355–361.

128   Berry, J.A. 1975. Adaptation of photosynthetic process to stress. *Science*, **188**, 644–650.

129   Chabot, B.F., Jurik, T.W. and Chabot, J.F. 1979. Influence of instantaneous and integrated light-flux density on leaf anatomy and photsynthesis. *Am. J. Bot.*, **66**, 940–945.

130   Cutter, E.G. 1971. *Plant anatomy: experiment and interpretation. Part 2 Organs.* London, Arnold.

131   Dale, J.E. 1976. Cell divisions in leaves. In: Yeoman, M.M. (ed.), *Cell Division In Higher Plants.* London, Academic Press, 315–345.

132   Turrell, F.M. 1936. The area of the internal exposed surface of dicotyledon leaves. *Am. J. Bot.*, **23**, 255–264.

133   Nobel, P.S. 1976. Photosynthetic rates of sun versus shade leaves of *Hyptis emoryi*. *Plant Physiol.*, **58**, 218–223.

134   Bjorkmann, O. 1981. Responses to different quantum flux densities. In: Lange, O.L., Nobel, C.B., Osmund, C.B. and Ziegler, H. (eds.), Physiological plant ecology. I. Responses to the physical environment. *Encycl. Plant Physiol.*, Berlin, Springer-Verlag. NS.12A, 57–107.

135   Boardmann, N.K. 1977. Comparative photosynthesis of sun and shade plants. *Ann. Rev. Plant Physiol.*, **28**, 355–377.

136   Vogel, S. 1968. Sun and shade leaves: differences in convective heat dissipation. *Ecol.*, **49**, 1203–1204.

137   McCain, D.C., Croxdale, J. and Markley, J.L. 1988. Water is allocated differently in chloroplasts in sun and shade leaves. *Plant Physiol.*, **86**, 16–18.

138   Kausch, W. and Haas, W. 1965. Chemische Unterschiedezwischen Sonnen-und Schattenblattern der blutbuche (*Fagus silvatica* cv. Atropunicea). *Naturwissenschaften*, **512**, 214–215.

139   Kasperbauer, M.J. 1971. Spectral distribution of light in a tobacco canopy and effects of end-of-day light quality on growth and development. *Plant Physiol.*, **47**, 775–778.

140   Sanchez, R. 1971. Phytochrome involvement in the control leaf shape of *Taraxacum officinale*. *Experimentia*, **27**, 1234–1237.

141   Satter, R.L. and Wetherell, D.F. 1968. Photomorphogenesis in *Sinningia speciosa* cv. Queen Victoria. II. Stem elongation: interaction of phytochrome controlled process and red light-requiring energy dependent reaction. *Plant Physiol.*, **43**, 961–967.

142   Cogliatti, D.H. and Sanchez, R.A. 1982. Influencia del fitocromo sobre el crecimiento foliar en *Taraxaum officinale*. *Phyton*, **42**, 191–199.

143   Casal, J.J., Aphalo, P.J. and Sanchez, R.A. 1987. Phytochrome effects on leaf growth and chlorophyll content in *Petunia axilaris*. *Plant Cell Environ.*, **10**, 509–514.

144   Van Volkenburgh, E., Hunt, S. and Davies, W.J. 1983. A simple instrument for the measuring cell wall extensibility. *Ann. Bot.*, **51**, 669–672.

145   Cleland, R.E. 1986. The role of hormones in wall loosening and plant growth. *Aust. J. Plant Physiol.*, **13**, 93–103.

146   Millthorpe, F.L. and Moorby, J. 1974. *An Introduction to Crop Physiology.* Cambridge, Univ. Press.

147   Friend, D.J.C. and Pomeroy, M.E. 1970. Changes in cell size and number associated with the effects of light intensity and temperature on the leaf morphology of wheat. *Can. J. Bot.*, **48**, 85–90.

148   Begg, J.E. and Wright, M.J. 1962. Growth and development of leaves from intercalary meristems in *Phalaris arundinacea. Nature*, **194**, 1097–1098.
149   Reid, D.M., Tuing, M.S. and Railton, I.D. Phytochrome control of gibberellin levels in barley leaves. *Plant growth substances 1973. Proc. VIII Int. Conf. Plant Growth Subst. Tokyo.* 325–331.
150   Casals, J.J., Sanchez, R.A. and Deregius, V.A. 1987. The effect of light quality on shoot extension growth in three species of grasses. *Ann. Bot.*, **59**, 1–7.
151   Casal, J.J. and Alvarez, M.A. 1988. Blue light effects on the growth of *Lolium multiflorum* leaves under natural radiation. *New Phytol.*, **109**, 41–45.
152   Bradbeer, J.W. 1973. The synthesis of chloroplast enzymes. In: Millborrow, B.V. (ed.), *Biosynthesis and its Control in Plants*. London, Academic Press, 279–302.
153   Bjorn, L.O. 1976. The state of protochlorophyll and chlorophyll in corn roots. *Physiol. Plant.*, **37**, 183–184.
154   Richter, G. and Dirks, W. 1978. Blue light induced development of chloroplasts in isolated seedling roots. Preferential synthesis of chloroplast ribosomal RNA species. *Photochem. Photobiol.*, **27**, 155–160.
155   Sundqvist, C., Bjorn, L.O. and Virgin, H.I. 1980. Factors in chloroplast differentiation. In: Reinert (ed.), *Chloroplasts*. Berlin, Springer-Verlag, 201–224.
156   Bjorkman, O., Boardman, N.K., Anderson, J.M., Thorne, S.W., Goodchild, D.J. and Pyliotis, N.A. 1972. Effect of light intensity during growth of *Atriplex patula* on the capacity of photosynthetic reactions, chloroplast, chloroplast components and structure. *Carnegie Inst. Wash. Year Book*, **71**, 115–135.
157   Mohr, H. 1984. Phytochrome and chloroplast development. In: Baker, N.R., Barber, J. (eds.), *Chloroplast Biogenesis*. Amsterdam, Elsevier, 307–343.
158   Biggins, J. 1987. *Progress in Photosynthesis Research*. Vol IV, Dordrecht, Martinus Nighoff.
159   Adamson, H., Griffiths, T., Packer, N. and Sunderland, M. 1985. Light-independent accumulation of chlorophyll-a and -b and protochlorophyllide in green barley *Hordeum vulgare. Physiol. Plant.*, **64**, 345–352.
160   Richter, G. 1969. Chloroplasyendifferenzierung in isolierten Wurzeln. *Planta.*, **86**, 299–300.
161   Bjorn, L.O. 1980. In: Senger, H. (ed.), *The Blue Light Syndrome*. Berlin Heidelberg New York, Springer Verlag, 455–646.
162   Koski, V.M., French, C.S. and Smith, J.H.C. 1951. The action spectrum for the transformation of protochlorophyll to chlorophyll-a in normal and albino corn seedlings. *Arch. Biochem. Biophys.*, 311–317.
163   Griffiths, W.T. 1978. Reconstitution of chlorophyllide formation by etioplasts membranes. *Biochem. J.*, **174**, 681–692.
164   Redlinger, T.E. and Apel, K. 1980. Light effects on protochlorophyllide-binding polypeptides. *Arch. Biochem. Biophys.*, **200**, 253–260.
165   Kasemir, H., Bergfeld, R. and Mohr, H. 1975. Phytochrome-mediated control of prolamellar body reorganisation and plastid size in mustard cotyledon. *Photochem. Photobiol.*, **21**, 111–120.
166   Girnth, C., Bergfeld, R. and Kasemir, H. 1978. Phytochrome-mediated control of grana and stroma thylakoid formation in plastids of mustard seedlings. *Planta.*, **141**, 191–198.
167   Usner, G. and Mohr, H. 1970. Phytochrome-mediated increase of galactolipids in mustard seedlings. *Naturwissenschaften*, **57**, 358–359.
168   Shibata, K. 1957. Spectroscopic studies on chlorophyll formation in intact leaves. *J. Biochem.*, **44**, 147–173.
169   Kasemir, H. 1983. Light control of chlorophyll accumulation in higher plants. In: Shropshire, W, and Mohr, H. (eds.), *Encyclopedia of Plant Physiol.*, NS. Vol 16B *Photomorphogenesis*. Berlin, Springer Verlag, 662–686.

170   Frosch, S., Bergfeld, R., Mehnert, C., Wagner, E. and Greppin, H. 1985. Ribulose bisphosphate carboxylase capacity and chlorophyll content in developing seedlings of *Chenopodium rubrum* growing under light of differential qualities and fluence rates. *Photosyn. Res.*, **7**, 41–67.

171   Tobin, E. and Silverthorne, J. 1985. Light regulation of gene expression in higher plants. *Ann. Rev. Plant Physiol.*, **36**, 569–593.

172   Kasemir, H., Rosemann, D. and Oelmuller, R. 1987. Changes in the appearance of ribulose bisphosphate carboxylase during senescence of mustard cotyledons. In: Biggins, J. (ed.), *Progress in Photosynthesis Research*. Dordrecht, Martinus Nighoff, IV 9, 561–564.

173   Anderson, J.M., Chow, W.S. and Goodchild, D.J. 1988. Thylakoid membrane organisation in sun/shade acclimation. *Aust. J. Plant Physiol.*, **15**, 11–26.

174   Anderson, J.M. 1986. Photoregulation of the composition, function and structure of thylakoid membranes. *Ann. Rev. Plant Physiol.*, **37**, 93–136.

175   Haehnel, W. 1984. Photosynthetic electron transport in higher plants. *Ann. Rev. Plant Physiol.*, **35**, 659–693.

176   Andersson, B. and Anderson, J.M. 1980. Lateral heterogeneity in the distribution of chlorophyll-protein complexes of the thylakoid membranes of spinach chloroplasts. *Biochem. Biophys. Acta*, **593**, 427–440.

177   Albertsson, P.A. 1971. *Partition of Cell Particles and Macromolecules*. 2nd ed. New York Wiley-Interscience.

178   Ackerlund, H.E., Andersson, B. and Albertsson, P.A. 1976. Isolation of photosystem II enriched membrane vesicles from spinach chloroplasts by phase partition. *Biochim. Biophys. Acta*, **449**, 525–533.

179   Miller, K.R. and Staehelin, L.A. 1976. Analysis of the thylakoid outer surface. Coupling factor is limited to unstacked membrane regions. *J. Cell Biol.*, **68**, 30–47.

180   Apek, K. 1979. The plastid membranes of barley (*Hordeum vulgare*). Light induced appearance of m-RNA coding for the apoprotein of the light harvesting chlorophyll a/b protein. *Eur. J. Biochem.*, **85**, 581–588.

181   Tobin, E.M. 1981. Phytochrome-mediated regulation of messenger RNAs for the small subunit of ribulose 1, 5 bisphosphate carboxylase, and the light harvesting chlorophyll a/b protein in *Lemna gibba*. *Plant Mol. Biol.*, **1**, 35–51.

182   Silverthorne, J. and Tobin, E.M. 1984. Demonstration of transcriptional regulation of specific genes by phytochrome action. *Proc. Nat. Acad. Sci.*, USA, **81**, 1112–1116.

183   Mosinger, E., Batschauer, A., Schafer, E. and Apel, K. 1985. Phytochrome control of *in vitro* transcription of specific genes in isolated nuclei from barley (*Hordeum vulgare*). *Eur. J. Biochem.*, **147**, 137–142.

184   Chow, W.S., Qian, L., Goodchild, D.J. and Anderson, J.M. 1988. Photosynthetic acclimation of *Alocasia macorrhiza* to growth irradiance: structure, function and composition of chloroplasts. *Aust. J. Plant Physiol.*, **15**, 107–122.

185   Terashima, I. and Inoue, Y. 1985. Vertical gradient in photosynthetic properties of spinach chloroplasts dependent on intra-leaf light environment. *Plant Cell Physiol.*, **26**, 781–785.

186   Lichtenhaler, H.K., Buschmann, C., Doll, M., Fietz, H.J., Bach, T., Kozel, U., Meier, D. and Rahmsdorf, U. 1981. Physiological activity, chloroplast ultrastructure and leaf characteristics of high light and low light plants and of sun and shade leaves. *Photosynthesis Res.*, **2**, 115–141.

187   Melis, A. and Harvey, G.W. 1981. Regulation of photosystem stoichiometry, chlorophyll-a and chlorophyll-b and relation to chloroplast ultrastucture. *Biochem. Biophys. Acta*, **637**, 138–145.

188   Glick, R.E., McCauley, S.W. and Melis, A. 1985. Effect of light quality on chloroplast membrane function in pea. *Planta.*, **164**, 487–494.

189 Eskins, K. and Duysens, M.E. 1984. Chloroplast structure in normal and pigment deficient soybean grown in continuous red and far-red light. *Physiol. Plant*, **61**, 351–356.

190 Eskins, K., McCarthy, S., Dyba, L. and Duysen, M. 1986. Corn chloroplast development in weak fluence rate red light and weak fluence rate red plus far-red light. *Physiol. Plant*, **67**, 242–246.

191 Oelmuller, R. and Mohr, H. 1985. Carotenoid composition in milo (*Sorghum vulgare*) shoots as effected by phytochrome and chlorophyll. *Planta.*, **164**, 390–395.

192 Leong, T.Y. and Anderson, J.M. 1983. Changes in composition and function of thylakoid membranes as a result of photosynthetic adaptation of chloroplasts from pea plants grown under different light conditions. *Biochim. Biophys. Acta*, **723**, 391–399.

193 Leong, T.Y., Goodchild, D.J. and Anderson, J.M. 1985. Effect of light quantity on the composition, function and structure of photosynthetic thylakoid membrane of *Asplenium australascium*. *Plant Physiol.*, **78**, 561–567.

194 Senger, H. and Bauer, B. 1987. The influence of light quality on adaptation and function of the photosynthetic apparatus. *Photochem. Photobiol.*, **45**, 936–946.

195 Voskresenskaya, N.P. 1980. Control of the activity of photosynthetic apparatus in higher plants. In: Senger, H. (ed.), *Blue Light Effects in Biological Systems*. Berlin, Springer-Verlag, 407–418.

196 Haupt, W. 1982. Light mediated movement of chloroplasts. *Ann. Rev. Plant Physiol.*, **33**, 205–233.

197 Kamiya, N. 1962. Protoplasmic streaming. *Enc. Plant Physiol.*, **172**, 979–1035.

198 Kamiya, N. 1981. Physical and chemical basis of cytoplasmic streaming. *Ann. Rev. Plant Physiol.*, **32**, 205–236.

199 Seitz, K. 1979. Cytoplasmic streaming and cyclosis of chloroplasts. *Enc. Plant Physiol.*, **7**, 150–169.

200 Zurzycki, J. 1955. Chloroplast arrangement as a factor in photosynthesis. *Acta Soc. Bot. Pol.*, **24**, 27–35.

201 Allen, N.S. and Allen, R.D. 1978. Cytoplasmic streaming in green plants. *Ann. Rev. Biophys. Bioeng.*, **7**, 497–526.

202 Williamson, R.E. 1980. Actin in motile and other processes in plant cells. *Can. J. Bot.*, **58**, 766–772.

203 Williamson, R.E. 1986. Organelle movements along action filaments and microtubules. *Plant. Physiol.*, **82**, 631–634.

204 Virgin, H.I. 1949. The relation between the viscosity of cytoplasm, plasma flow, and the motive force. *Physiol. Plant*, **2**, 157–163.

205 Seitz, K. 1971. Die Ursache der Phototaxis der chloroplasten: ein ATP-gradient? *Z. Pflanzenphysiol.*, **64**, 245–256.

206 Seitz, K. 1967. Wirkungsspektren fur die Starklichtbewegung der Chloroplasten die Photodinese und die lichtabhangige viskostatsanderung bei *Vallisneria spiralis*. *Z. Pflanzenphysiol.*, **56**, 246–261.

207 Seitz, K. 1979b. Light induced changes in the centrifugability of chloroplasts: different action spectra and different influence of inhibitors in the low and high intensity range. *Z. Pflanzenphysiol.*, **95**, 1–12.

208 Zurzycki, J. 1967. Properties and localisation of the photoreceptor active in the displacement of chloroplasts in *Funaria hygrometrica*. *Acta Soc. Bot. Pol.*, **36**, 143–152.

209 Satter, R.L. 1979. Leaf movements and tendril curling. In: Haupt, W. and Feinlieb, M.E. (eds.), *Physiology of Movement. Encyclopedia of Plant Physiol.* NS. **7**, 442–484.

210 Satter, R.L. and Galston, A.W. 1981. Mechanism of control of leaf movement. *Ann. Rev. Plant Physiol.*, **32**, 83–110.

211 Koukari, W.L. and Hillman, W.S. 1968. Pulvini as the photoreceptors in the

phytochrome effect on nyctinasty in *Albizzia julibrissin. Plant Physiol.*, **43**, 698–704.

212   Satter, R.L. and Galston, A.W. 1971a. Potassium flux: a common feature of *Albizzia* leaflet movement controlled by phytochrome or endogenous rhythm. *Science*, **174**, 518–520.

213   Satter, R.L. and Galston, A.W. 1971b. Phytochrome-controlled nyctinasty in *Albizzia julibrissin*. III. Interaction between an endogenous rhythm and phytochrome in control of potassium flux in leaflet movement. *Plant Physiol.*, **48**, 740–746.

214   Campbell, N.A. and Garber, R.C. 1980. Vacuolar reorganization in the motor cells of *Albizzia julibrissin* during leaf movement. *Planta.*, **148**, 251–255.

215   Bunning, E. and Moser, I. 1969. Response-kurven bei der Circadianen rhythmik von *Phaseolus. Planta.*, **69**, 101–110

216   Bjorkman, O. and Powles, S.B. 1981. Leaf movement in the shade species *Oxalis oregana*. I. Response to light level and light quality. *Carnegie Inst. Washington Year Book.* **80**, 59–62.

217   Powles, S.B. and Bjorkman, O. 1981. Leaf movement in the shade species *Oxalis oregana*. II. Role in protection against injury by intense light. *Carnegie Inst. Washington Year Book*, **80**, 63–66.

218   Koller, D. 1986. The control of leaf orientation by light. *Photochem. Photobiol.*, **44**, 819–826.

219   Sheriff, D.W. and Ludlow, M.M. 1985. Diaheliotropic responses of leaves of *Macroptilium atropurpureum*. cv. Siratro. *Aust. J. Plant Physiol.*, **12**, 151–171.

220   Schwartz, A. and Koller, D. 1986. Diurnal phototropism in the solar tracking leaves of *Lavatera cretica. Plant Physiol.*, **80**, 778–781.

221   Volgelmann, T.C. and Bjorn, L.O. 1983. Responses to directional light by leaves of a sun tracking lupine (*Lupinus succulentus*). *Physiol. Plant*, **59**, 533–538.

222   Schwartz, A., Gibola, S. and Koller, D. 1987. Photonastic control of leaflet orientation in *Melilotus indicus* (Fabacae). *Plant Physiol.*, **84**, 318–323.

223   Wooley, J.T., Alfrich, R.A. and Larson, E.M. 1984. Direct measurement of irradiance upon phototropic soybean leaves throughout the day. *Crop Sci.*, **24**, 614–616.

224   Rajendrudu, G. and Rama Das, V.S. 1981. Solar tracking and light interception by leaves of some dicot species. *Current Sci.*, **50**, 618–620.

225   Wilmer, C.M. 1983. *Stomata*. London, Longman.

226   Ting, I.P. 1985. Crassulacean acid metabolism. *Ann. Rev. Plant Physiol.*, **36**, 595–622.

227   Hsiao, T.C., Allaway, W.G. and Evans, L.T. 1973. Action spectra for guard cell Rb^+ uptake and stomatal opening in *Vicia faba. Plant Physiol.*, **51**, 82–88.

228   Sharkey, T.D. and Raschke, K. 1981a. Separation and measurement of direct and indirect effects of light on stomata. *Plant Physiol.*, **68**, 53–40.

229   Schwartz, A. and Zeiger, E. 1984. Metabolic energy for stomatal opening. Roles of photophosphorylation and oxidative phosphorylation. *Plant Physiol.*, **161**, 129–136.

230   Zeiger, E. 1986. The photobiology of stomatal movements. In: Kendrick, R.E. and Kronenberg, G.H.M. (eds.), *Photomorphogenesis in plants*. Dordrecht, Martinus Nijhoff, 391–413.

231   Zeiger, E., Assmann, S.M. and Meidner, H. 1983. The photobiology of *Paphiopedilum* stomata; opening under blue but not red light. *Photochem. Photobiol.*, **37**, 627–630.

232   Roth-Bejerano, Nejidat, A. 1987. Phytochrome effects on K^+ fluxes in guard cells of *Commelina communis. Physiol. Plant*, **71**, 345–351.

233   Zeiger, E., Armond, P. and Melis, A. 1981a. Fluorescence properties of guard cell chloroplasts. *Plant Physiol.*, **67**, 17–20.

234 Zeiger, E. and Hepler, P.K. 1977. Light and stomatal function: blue light stimulates swelling of guard cell protoplasts. *Science*, **196**, 887–889.

235 Zeiger, E. and Field, C. 1981. Photocontrol of the functional coupling between photosynthesis and stomatal conductance in the intact leaf. *Plant Physiol.*, **70**, 370–375.

236 Zeiger, E., Field, C. and Mooney, H.A. 1981b. Stomatal opening at dawn: possible roles of the blue light response in nature. In: Smith, H. (ed.), *Plants and the Daylight Spectrum*. New York, Academic Press, 391–407.

237 Goldberg, B. and Klein, W.H. 1977. Variations in the spectral distribution of daylight at various geographical locations on the earth's surface. *Solar Energy*, **19**, 3–13.

238 Zeiger, E., Bloom, A.J. and Hepler, P.K. 1978. Ion transport in stomatal guard cells. A chemiosmotic hypothesis. *Whats New in Plant Physiology?* **9**, 29–32.

239 Holmes, M.G. and Klein, W.H. 1986. Photocontrol of dark circadian rhythms in stomata of *Phaseolus vulgaris*. *Plant Physiol.*, **82**, 28–33.

240 Osborne, D.J., Jackson, M.B. and Millborrow, B.V. 1972. Physiological properties of abscission accellerator from senescent leaves. *Nature* (New Biology), **240**, 98–101.

241 Osborne, D.J. 1973. Internal factors regulating abscission. In: Kozlowski, T.T. (ed.), *Shedding Plant Parts*. New York, Academic Press, 125–148.

242 Addicott, F.T. and Lyon, J.L. 1973. Physiological ecology of abscission. In: Kozlowski, T.T. (ed.), *Shedding Plant Parts*. New York, Academic Press, 85–117.

243 Matzke, E.B. 1936. The effect of street lights in delaying leaf fall. *Am. J. Bot.*, **23**, 446–452.

244 Curtis, R.W. 1978a. Induction of resistance to dark abscission by malformin in white light. *Plant Physiol.*, **62**, 264–266.

245 Curtis, R.W. 1978b. Participation of phytochrome in the light inhibition of malformin-induced abscission. *Plant Cell Physiol.*, **19**, 289–297.

246 Curtis, R.W. 1978c. Phytochrome involvement in the induction of resistance to dark abscission by malformin. *Planta*, **141**, 311–314.

247 Decoteau, D.R. and Craker, L.E. 1983. Abscission: Quantification of light control. *Plant Physiol.*, **73**, 450–451.

248 Decoteau, D.R. and Craker, L.E. 1984. Abscission: Characterisation of control. *Plant Physiol.*, **75**, 87–89.

249 Decoteau, D.R. and Craker, L.E. 1987. Abscission: Ethylene and light control. *Plant Physiol.*, **83**, 970–972.

250 Craker, L.E., Zhao, S.Y. and Decoteau, D.R. 1987. Abscission: Response to red and far-red light. *J. Exp. Bot.*, **38**, 883–888.

6
The Flowering Process

Introduction

Flowering in many plants is a photoperiodic phenomenon, that is, a response which is controlled by the duration of light. Although flowering is the best known and most investigated photoperiodic effect in plants, it is by no means the only one. Photoperiodic effects are mentioned in seed germination (p. 49), leaf abscission (p. 101), bulbing and tuber formation (p. 135) and are also reported for rooting[1] and runner development.[2] Responsiveness to photoperiod can form an effective strategy. It allows synchronization between the life cycle of the plant and seasonally associated changes in the environment. Flowering in long-day plants (LDP) at high latitudes is associated with a period of the year which can supply high fluence rates to meet the energy requirement for seed formation. A short-day (SD) response at a low latitude might serve to synchronize development with the end of a tropical rainy season.

The twentieth-century fathers of photoperiodic studies are generally regarded to be Garner and Allard (1920, 1922).[3,4] These workers were plant breeders interested in getting flowers and seeds from two species, the soy bean, *Glycine max* var. Biloxi, and a newly discovered variety of *Nicotiana tabacum* called Maryland Mammoth. They observed that under natural conditions, Biloxi would flower in September or October regardless of whether the seeds were germinated in May, June, July or August. Maryland Mammoth, on the other hand, failed to flower in the field and needed to be brought into the greenhouse and propagated. Quite small plants obtained in this way flowered profusely in the subsequent winter. Garner and Allard were soon able to show that it was daylength, among the environmental variables, which governed flowering in both these species. Although the contribution of these workers to the understanding of flowering is outstanding, several earlier workers should also be noted. Tournois (1914)[5], from his studies of hop (*Houblon japonais*) and marijuana or hemp (*Cannabis sativa*), concluded that short durations of light were required to promote flowering in these species and that it was the length of the night period rather than the length of the light period which was the most important factor in the diurnal cycle. Klebs (1918)[6] provided evidence from his experiments with the house leek (*Sempervivum funkii*) that the flowering response was controlled by the duration of light rather than the fluence level.

Response types

Early work led to the division of plants into three main classes. These are long-day plants (LDP), short-day plants (SDP) and day-neutral plants (DNP). LDP may be defined as plants which flower or show accelerated flowering as a result of an increase in daylength beyond a critical value and obviously SDP flowers, or shows accelerated flowering, with a shortening of the photoperiod below a critical value. A DNP is a plant which flowers independent of the photoperiod. It is important to realise that the actual duration of light is not critical to these definitions. *Xanthium pennsylvanicum* is a SDP which has been widely used in the study of flowering and has a critical daylength of 15.75 h. That the photoperiod is nearly 66 per cent of the total cycle does not detract from *Xanthium* being a SDP. The salient point is that when the daylength is reduced below this period, flowering is promoted. If a generalisation is to be made, it is that SDP are plants native to low latitudes, whereas LDP originate in higher latitudes.

As more plants have been studied, it has been necessary to widen the classification of flowering responses. Plants are now known which have dual daylength requirements. For example, *Cestrum nocturnum* requires LD followed by SD and flowers in the autumn, whereas *Scabiosa succisa*, which requires SD followed by LD, flowers in the spring. A lesser known group of plants are those which have intermediate daylength requirements responding to a restricted range of photoperiods when days are not too long and not too short.

Flowering and experimentation

The flowering process can be divided into two stages: the initiation of flowers and the development of these into open blooms. These two processes may have different daylength requirements. *Chrysanthemum morifolium* is a SDP which develops flower buds when the days shorten below 14.5 h but rapid development of the buds depends upon a further shortening of the daylength to 13 to 13.5 h. This obviously suits autumn flowering.[7]

The initiation of flower buds requires vegetative apices producing leaf primordia to be transformed into flowering apices producing flower parts. Plants are frequently selected for experimentation because they are sensitive to a single inductive cycle (i.e. appropriate daylength to cause flowering), e.g. *Xanthium* and *Pharbitis*. Whereas simple systems are always attractive to the investigator, simple requirements may not be typical of flowering plants. Experimental material is also selected for ease of examining the apices. It is rare that experiments proceed to the open flower stage and plants are often scored as flowering when the initial anatomical changes have taken place at the apex. These are normally detected with nothing more elaborate than a hand lens. In some experiments where qualitative differences are thought to be relevant, the flower development is arbitrarily divided into stages, and stage of development is scored.

Juvenility

Plants are unable to respond to the appropriate photoperiod when still in the juvenile stage. The juvenile stage may last from a few days in herbaceous plants to a few years in woody species. Evidence exists to support both the hypothesis that there is a plant-apex signal which brings about transition to the mature phase and vice versa. Nothing is known about the factors that influence the plant to develop sensitivity to photoperiod.

Photoperception

Julius Sachs (1880) suggested that leaves produce small quantities of substances which directly assimilate into flower formation, but it was not until the work of Chailakhyan (1936)[8] that real progress was made in terms of identifying the site of photoperception. Working with *Chrysanthemum*, Chailakhyan demonstrated that flowering was induced when the whole plant was exposed to SD. If the leaves were exposed to SD while the apex was exposed to LD, the plants still flowered. If the leaves were exposed to LD and the apex to SD, flowering did not ensue. From these experiments, Chailakhyan suggested that the leaf was the site of perception of the inductive photoperiod and synthesised a flowering hormone which he called florigen. He suggested that this substance was transmitted from the site of synthesis in the leaf to the apex, where it induced flowering. Hamner and Bonner went on to show[9] that if a single leaf of *Xanthium* (SDP) was exposed to the appropriate photoperiod while the rest of the plant remained in long days, then the plant was induced to flower. If a single leaf of either *Xanthium* or *Perilla* (SDP) was removed from a plant which had been given inductive daylengths and grafted on to a plant kept under non-inductive conditions, then the presence of the grafted leaf induced flowering.[10] With *Perilla* the grafted leaf remained in good condition and was removed and grafted on to another plant kept under non-inductive conditions, where it also caused flowering, 3 months after it was last exposed to SD.

It has been known for many years that the floral stimulus travels in the phloem to the apex. Any treatment which interrupts the continuity of this tissue inhibits or delays flowering. Where the floral stimulus enters the phloem is not clear. In the lamina of the leaf, the stimulus seems to be capable of travelling through the epidermis and/or mesophyll since cutting the veins approaching the mid-rib does not seem to have a marked effect on flowering. By the time the flowering stimulus reaches the petiole, it can be established that it is travelling in the phloem. Experiments by a number of workers to determine whether the floral stimulus travels passively with the carbohydrate or whether it has a specific transport mechanism within the phloem have led to a degree of confusion. These experiments involve defoliation of the experimental plant with the exception of a single leaf. This is supplied with $^{14}CO_2$ and the appropriate photoperiod simultaneously. The apex is then monitored for the arrival of radioactive assimilates and the first sign of floral initiation. In some species these events occur together while in others one or the other occurs first. The separation of these two events in time varies from species to species.

Whether the leaf is the site of photoperception in the gymnosperms is a matter of doubt. The scales around the buds containing the female strobiles of *Picea abies* and *Pinus sylvestris* absorb more R than FR.[11] In consequence it may be predicted that the apical domes of the buds contain little Pfr. If under natural daylight optical fibres are pushed into these buds and left in this condition for a year, the light which enters via the optical fibre would certainly increase the Pfr/Ptot inside the bud, and it is suggested that this variable causes a marked increase in flowering.

A universal flowering hormone?

From studies with inter-specific graft unions it may be suggested that there is evidence for a universal flowering hormone. By grafting the SDP *Xanthium* to the LDP *Rudbeckia bicolor* and keeping the plants under LD, the *Xanthium* could be induced to flower.[12] If *Nicotiana tabacum* (SDP) is held under SD while grafted to *Hyoscyamus niger* (LDP), then the latter will be induced to flower.[13]

Hormone purification

Isolation of a flowering hormone which might cause a variety of species to flower when applied exogenously has aroused the interest of a number of laboratories over the years. These attempts have been largely unsuccessful. One of the more exciting results was reported by Lincoln *et al.*[14] These workers made freeze-dried extracts of flowering *Xanthium* plants in absolute methanol. These extracts were incorporated into lanolin and applied to the leaves of *Xanthium* under LD. The extract was also active when supplied via a stem flap. Between 10 and 13 per cent of the plants reached an early stage of flowering and some were recorded as reaching anthesis. Extracts from *Xanthium* kept under LD were not active. A purification of the active extract was attempted and it was reported to have the partitioning properties of a water-soluble carboxylic acid. Unfortunately further purification resulted in a loss of flower-promoting activity. These results have been confirmed, and gibberillic acid (GA_3) was added to the extract.[15] This considerably enhanced the response; 50 per cent of plants were reported as flowering. No female flowers ever developed and at first the male inflorescences which appeared were deformed, although normal male flowers did develop after some months.

Growth substances and flowering

The failure of many attempts to identify florigen has led to the suggestion that it may be a combination of hormones. IAA, ethylene, cytokinins and ABA can all be shown to affect flowering in isolated cases but endogenous gibberillic acid (GA) can be shown to be regulated by daylength and is therefore favoured to be involved with flowering. Exogenously applied GA is known to cause flowering in many LDP and a few SDP in non-inductive conditions. GA will also inhibit flowering in many SDP and a few LDP. GA is particularly effective in rosette LDP where stem extension and flowering are anatomically associated but there are exceptions to this rule. The stem extension response in the LDP *Silene* is inhibited by AMO1618 (an inhibitor of GA synthesis) but flowering is unaffected. However, inhibitors of GA synthesis and action suppress flowering in *Pharbitis* (SDP). [16]

Chailakhyan (1975)[17] has suggested that the flowering stimulus is a combination of GA and other hormones, which he calls anthesins. He suggests that LDP have a high level of anthesins and that GA limits the flowering process, while in SDP the GA level is high and the anthesins are limiting. His group was able to extract three GA-like substances from the LDP *Nicotiana sylvestris* when kept under LD, which could cause flowering in the LDP *Rudbeckia* kept under SD. [18] The GA-like substances which were extracted from *N. sylvestris* under SD were ineffective in producing flowers in *Rudbeckia*.

When LDP cultivars of *Pisum sativum* were kept under SD and supplied with tritiated GA_9, it was metabolised into more polar GA. It has been suggested that polar GA produced in LDP by SD inhibit flowering, whereas in LD its formation is blocked. [19]

The importance and relationship of dark and light periods

In the early days of research into flowering, the question as to whether the plant was responding to the light period or the dark period arose. This question was first approached by Hamner (1940). [20] Using the SDP *Glycine max*, he varied the light period while keeping the dark period constant and *vice versa*. He was able to show that the ratio of light to dark was not relevant to flowering nor was the length of the light period critical. A dark period of 10 h was shown to promote flowering whether accompanied by 4 or 16 h of light. From this work it was concluded that there was a critical night length that must be exceeded to induce SDP to flower. Since this time, flowering has been examined in a large number of species, and the overriding feature is that the night length is of critical importance in flowering.

The dark period

An outstanding feature of the dark period, particularly in SDP, is that it must be uninterrupted. If a short period of light is supplied (night break) during a dark period of inductive length the response is nullified. This phenomenon led to the elucidation of the phytochrome involvement with photoperiodic effects. It was demonstrated that the most effective wavelengths to use as a night break were R and that this effect was, to a reasonable degree, R/FR reversible in both *Xanthium* and *Pharbitis*. [21,22] The effectiveness of the night break varies with the time it is given after the beginning of the inductive dark period. [23,24] From this work it becomes apparent that the phytochrome system interacts with an endogenous rhythm (see below). If the involvement of phytochrome in LDP is related to that in SDP, then night breaks with R should promote flowering. A few such examples can be found, e.g. *Hyoscyamus*, [25] but many LDP are not sensitive to night breaks. Action spectra of night breaks are remarkably similar in both the LDP *Hyoscyamus niger* and *Hordeum vulgare* and the SDP *Xanthium pennsylvanicum* and *Glycine max*, although they have the opposite effects. [21]

Measurement of the dark period

Two basic hypotheses have been forwarded to explain the measurement of time in photoperiodic phenomena. First is the hour glass hypothesis, which suggests that biochemical reactions are initiated by the onset of darkness. If these reactions proceed uninterrupted by light, then the products of these reactions accumulate and induce flowering. The Pfr form of phytochrome decays and reverts at predictable rates and at first sight would form an hour glass system for measuring the length of the dark period.[26] In SDP, the dark period would need to be long enough to allow the high level of Pfr that may be present at the beginning of the dark period to decay or revert to Pr and allow flowering. In other words, high levels of Pfr throughout the dark period inhibit flowering in SDP. When this idea is extrapolated to LDP, it is suggested that the dark period needs to be short enough to allow Pfr to remain above a certain high level throughout the dark period. However, this simplistic approach does not fit all known data. First, the temperature sensitivity of reversion and decay rates are not reflected in flowering, and such a temperature-sensitive system would make an inaccurate hour glass. Second, the rhythmic nature of light interruption responses in prolonged dark period (p. 127) would be difficult to explain in terms of a Pfr hour glass. The phytochrome content of *Pharbitis nil*, grown in various ways to avoid chlorophyll formation, has been examined and it is reported that there is no correlation between the form or level of spectrophotometrically measurable phytochrome and the night break sensitivity.[27]

The second hypothesis meets with greater favour. It suggests that dark timing is related to the progress of a circadian rhythm. Cumming *et al.* (1965)[23] show that if *Chenopodium rubrum* (SDP) is subjected to prolonged dark periods (about 100 h) and the sensitivity of the plants to R breaks is measured throughout this period, the pattern that emerges is a rhythm with a period of approximately 30 h. This begins to damp out over the experimental period (see Figure 7.2). Although phytochrome itself is not the timing mechanism, it must be closely connected with it since there is evidence from action spectra that suggests phytochrome provides the flowering response with the dark-light transition signal. This concept is discussed in the next chapter.

The light period

Flowering involves localised growth, and both the energy and the building blocks for this process originate from photosynthesis. However, the involvement of photosynthesis and flowering is by no means clear. Experiments carried out with *Chrysanthemum* suggest that it is contemporary photosynthesis which contributes to the flowering process and photosynthetic history of the plant is largely irrelevant.[28] On the other hand, a direct link between these two processes is not apparent in other plants. *Kalanchoe* (LDP) can be induced to flower when given very short, intense periods of light considered to be of no photosynthetic consequence.[29] Even more perplexing is the response of young dark-grown *Pharbitis nil* (SDP), which can be induced to flower if an R pulse and benzyladenine (a cytokinin) are supplied at the same time.[30]

Vernalisation

Interaction of light and temperature have been noted earlier in seed germination (p. 59) and these phenomena may well be related to vernalisation. Vernalisation may be generally defined as the requirement of a plant for a low temperature treatment which enhances flowering. The temperatures most commonly required are between 1 and 7°. More specifically, for historical reasons, the term is used to describe the promotion of flowering in winter varieties of cereals by cold treatment of moistened or germinating seeds. Winter varieties are those cereals which are sown in the autumn when the soil is suitable for cultivation and which develop slowly until halted by low soil temperatures.

Temperature can have direct effects upon flowering in some plants but these may be separated from vernalisation since vernalisation is an inductive phenomenon. That is to say, when the vernalisation treatment is completed, flower initials are not yet present, whereas direct effects of temperature take place after the initiation of flowering apices. Vernalisation is to some degree reversible. The most common treatment which will de-vernalise plants is high temperature, but it is also known to happen as a result of SD[31] and there may be ecological significance to this finding.

Vernalisation is a common requirement for flowering in many LDP and less common among SDP. It occurs in annuals (*Triticum* and *Pisum*), in biennials (*Daucus* and *Hyosycamus*) and perennials (*Chrysanthemum* and *Lolium*). Experiments in which different parts of the plant were given localised low temperature treatments[32] show that it is the stem apex which perceives the vernalising treatment. This contrasts with the perception of photoperiod for flowering. A further difference between these phenomena is that, in the majority of species studied, the vernalised plant cannot transmit a stimulus via a graft union to a non-vernalised plant. Metabolic differences have been demonstrated between vernalised and non-vernalised apices.[33] Using winter wheat embryos maintained on sucrose, vernalising treatment was found to induce the appearance of new proteins. Gel patterns of protein populations of vernalised winter wheat showed similarities to that of spring wheat which did not appear in the protein populations of non-vernalised winter wheat. The biennial strains of *Melilotus alba* and *M. officinalis* could be made to flower in their first year if exposed to a large number of long photoperiods.[34] These plants, however, did not develop the characteristic tap root by which the plant normally overwinters. The development of the tap root is initiated by SD. When the flowering capacities of plants derived from vernalised and non-vernalised tap roots were compared it was found that the vernalised plants flowered rapidly and required only 14 h photoperiods, whereas the non-vernalised plants required 17 h photoperiods to flower. These plants seem to modify their life cycle according to latitude. The perennial habit is the norm at lat. 42°N, whereas in Alaska at lat. 61°N, where the maximum daylength is 19.5 h and midsummer twilight lasts throughout the night, both *M. albus* and *M. officialis* are able to flower freely in the first year of growth.

Moonlight effects on flowering

Whether moonlight has sufficient energy to be detected by plants and is used as an environmental signal has been discussed earlier (p. 20). It is, however, worthy of reconsideration in relation to photoperiodic phenomena. Bunning and Moser (1969)[35] have suggested that the mechanism controlling sleep movements in legumes has been selected to ignore the stimulation of moonlight. This avoids the broad, light-absorbing surface of the leaf being at right angles to the incident radiation (moonlight) at night. Earlier workers had found that moonlight was capable of influencing flowering in species which did not display circadian leaf movements. If at night-time the LDP *Hordeum distichon* and *Triticum vulgare* were exposed to moonlight filtered through glass throughout the 8-week period of development, the flowering of these plants was advanced by 2 to 3 days compared with that of control plants, which were covered by a black cloth at night.[36] Of the SDP investigated, it was found that flowering in *Pharbitis hispida* and *Soja hispida* could be delayed by exposure to moonlight by 3 to 3.5 days. Results with one other LDP and two SDP were not always clear cut. This phenomenon has been investigated further using *Pharbitis nil* (SDP).[37] This plant may be induced to flower by a single inductive cycle and is a good experimental subject since it is particularly sensitive to night break inhibition of flowering. It was concluded that moonlight is capable of a slight delaying effect on flowering, but in the natural environment the effect of night temperature on flowering time is of greater significance.

References for Chapter 6

1　Langton, F.A. 1977. Genetic variation and photoperiodic response in *Chrysanthemum. Scientia Hortic.*, **7**, 277–289.

2　Guttridge, C.G. 1960. Photoperiodic responses in *Fragaria. Bull. L'Inst. Agron. Stat. Rech. Gembloux, hors serie.*, **2**, 941–948.

3　Garner, W.W. and Allard, H.A. 1920. Effects of the relative length of the day and night and other factors of the environment on growth and reproduction in plants. *J. Agric. Res.*, **18**, 553–603.

4　Garner, W.W. and Allard, H.A. 1922. Further studies in photoperiodism, the response of the plant to the relative length of day and night. *J. Agric. Res.*, **23**, 871–920.

5　Tounois, J. 1914. Etudes sur la sexualité du houblon. *Ann. Sci. Nat. Bot.*, **19**, 49–191.

6　Klebs, H. 1918. Uber die Blutenbidung bei *Sempervivum. Flora*, (Jena) **128**, 111/112.

7　Cathey, H.M. 1969. *Chrysanthemum morifolium*. In: Evans, L.T. (ed.), *The Induction of Flowering*. Ithaca, Cornell Univ. Press, 268–290.

8　Chailakhyan, M.Kh. 1936. On the hormonal theory of plant development. *Dokl. Acad. Sci. USSR*, **12**, 443–447.

9　Hamner, K.C. and Bonner, J. 1938. Photoperiodism in relation to hormones as factors in floral initiation. *Bot. Gaz.*, **100**, 388–431.

10　Zeevaart, J.A.D. 1958. Flower formation as studied by grafting. *Meded Landb-Hoogesch Wageningen*, **58**, 1–88.

11   Kosinski, G. and Giertych, M. 1982. Light conditions inside developing buds affect floral induction. *Planta.*, **155**, 93–94.

12   Okuda, M. 1953. Flower formation of *Xanthium canadense* under long day conditions induced by grafting with long day plants. *Planta.*, **66**, 247–255.

13   Lang, A. and Melchers, G. 1948. Die photoperiodische Reaction von *Hyoscyamus niger*. *Planta.*, **33**, 654–702.

14   Lincoln, R.G., Mayfield, D.L. and Cunningham, A. 1961. Preparation of a floral initiating extract from *Xanthium*. *Science*, **133**, 756–760.

15   Carr, D.J. 1967. The relationship between florigen and the flowering hormone. In: Fredrick, J.F. and Weyer, E.M. (eds.), *Plant Growth Regulators*, New York, *Ann. Acad. Sci.*, **144**, 305–312.

16   Cleland, C.F. and Zeevaart, J.A.D. 1970. Gibberellins in relation to flowering and stem elongation in the long day plant. *Plant Physiol.*, **46**, 392–400.

17   Chailakhyan, M.Kh. 1975. Substances of plant flowering. *Biol. Plant*, **71**, 1–11.

18   Grigorieva, N.Y., Kucherov, V.F., Lozhnikova, V.N. and Chailakhyan, M.Kh. 1971. Endogenous gibberellin-like substances in long day species of tobacco—a possible correlation with photoperiodic response. *Phytochem.*, **10**, 509–517.

19   Proebsting, W. and Heftman, E. 1980. The relationship of (^{3}H) GA$_9$ metabolism to photoperiod-induced flowering in *Pisum sativum*. *Z. Planzen Physiol.*, **98**, 305–309.

20   Hamner, K.C. 1940. Interrelationship of light and darkness in photoperiodic induction. *Bot. Gaz.*, **101**, 658–687.

21   Borthwick, H.A., Hendrick, S.B. and Parker, M.V. 1948. Action spectrum for the photoperiodic control of floral induction of a long day plant Wintex Barley (*Hordeum vulgare*). *Bot. Gaz.*, **110**, 103–108.

22   Downs, R.J. 1956. Photoreversibility of flower initiation. *Plant Physiol.*, **31**, 278–284.

23   Cumming, B.G., Hendricks, S.B. and Borthwick, H.A. 1965. Rhythmic flowering responses and phytochrome changes in a selection of *Chenopodium rubrum*. *Can. J. Bot.*, **43**, 825–853.

24   Salisbury, F.B. and Bonner, J. 1956. The reactions of the photo-inductive dark period. *Plant Physiol.*, **31**, 141–147.

25   Claes, H. and Lang, A. 1947. Die Wirkung von Indoley lessigaure und 2,3,5 Triodbenzoesue auf Bluten Bildung Von *Hyosycamus niger*. *Zeitsshrift fur Naturschung*, **26**, 56–63.

26   Hendricks, S.B. 1960. Rates of change of phytochrome as an essential factor determining photoperiodism in plants. *Cold Spring Harbour Symp. Quant. Biol.*, **25**, 245–248.

27   King, R.W., Vince-Prue, D. and Quail, P. 1978. Light requirement, phytochrome and photoperiodic induction of flowering in *Pharbitis nil*. III. *Planta.*, **141**, 15–22.

28   Vince, D. 1960. Low temperature on the flowering of *Chrysanthemum morifolium*. *J. Hort. Sci.*, **35**, 161–175.

29   Schwarbe, W.W. 1959. Studies of long day inhibition in short day plants. *J. Exp. Bot.*, **10**, 317–329.

30   Ogawa, Y. and King, R.W. 1979. Establishment of photoperiodic sensitivity by Benzyladenine and a brief red irradiation of dark grown seedlings of *Pharbitis nil*. *Plant Cell Environ.*, **20**, 119–122.

31   Wellensiek, S.J. 1965. Recent development in vernalisation. *Acta Bot. Neerl.*, **14**, 308–314.

32   Purvis, O.N. 1940. Vernalization of fragments of embryo tissue. *Nature*, **145**, 462.

33   Teroaka, H. 1972. Proteins of wheat embryos in the period of vernalisation. *Plant Cell Physiol.*, **8**, 87–95.

34 Kasperbauer, M.J., Gardner, F.P. and Loomis, W.E. 1962. Interaction of photoperiod and vernalisation in flowering of sweet clover (*Melilotus*). *Plant Physiol.*, **37**, 165–170.

35 Bunning, E. and Moser, I. 1969. Interference of moonlight with the photoperiodic measurement of time by plants and their adaptive reaction. *Proc. Nat. Acad. Sci. USA*, **62**, 1018–1022.

36 Gaertner, T. and Braunroth, E. 1935. Uber den Einfdluss des Mondlichtes auf die Bluhtermin der Lang- und Kruztagspflanzen. *Botanisches Zentralblatt.* (*Abt. A*)., **53**, 554–563.

37 Kadman-Zahavi, A. and Peiper, D. 1987. Effects of moonlight on flower induction in *Pharbitis nil* using a single dark period. *Annals of Bot.*, **60**, 621–623.

7

Endogenous Rhythms

Introduction

In previous chapters the environmental influence on and the mechanism of stomatal movements, flowering and leaf orientation have been discussed. These phenomena are governed by endogenous rhythms and although they may share a common mechanism, they function to adapt the plant to different aspects of the daily or yearly cycle. Other plant processes which are influenced by rhythms include petal movement, photosynthesis, dark CO_2 fixation, respiration and root exudation.

Some plant rhythms adjust the plant to daily fluxes in temperature and light. This type of rhythm is known as circadian (circa = about, dien = day) and has a period, which is the time between repeating parts of the rhythm, of about 24 h. These rhythms differ from periodic responses in that the rhythm will persist to some degree when removed from external stimulation. The rhythm in CO_2 output found in *Bryophyllum fedtschenkoi* damps out slowly over 5 or 6 cycles in continuous darkness.[1] Major changes occur on a daily basis in the environment of the plant and it may well be a strategic advantage to the plant to be able to anticipate these events. The adaptive value of some manifestations of the biological clock is not immediately obvious. Circadian leaf movements, for example, have perplexed scientists for generations, but the suggestion finally came from Bunning (who admits the idea came to him in a dream)[2] that it is to hide the leaf surface from the influence of moonlight, which might interfere with photoperiodism.

Photoperiodism in plants is normally associated with flowering but also affects other aspects of development and confers on a plant the ability to occupy a seasonal niche. In other words, the plant is able to orientate its life cycle around particular times of the year.

To be useful, a plant's biological clock needs to be accurate in anticipating changes such as temperature and fluence. This may be achieved either by the accurate detection of time or, more likely, by the rapid resetting of the clock by environmental influences. Despite this sensitivity, the clock must be unaffected by transient variations in the environment.

The influence of light on circadian rhythms

It became apparent during early research on this subject[3] that when circadian rhythms such as the leaf movements in *Phaseolus multiflorus* were allowed to run free in continuous darkness, the free-running period is longer than 24 h. In the natural environment rhythms must be entrained or adjusted to the 24 h cycle. Contradicting the current dogma of the day, Bunning and Stern (1930) suggested[3] that the rhythm could be reset by exposure of the plants to low-fluence R. Lorcher (1957) produced a rough action spectrum for the initiation of leaf movements in dark-grown bean plants.[4] He found that the most effective wavelengths were R and that the stimulatory effects of R could be reversed by FR. Entrainment of this rhythm in light-grown plants is sensitive to R but other wavelengths are also effective. A more precise action spectrum has been supplied for a phase shift of the rhythm of CO_2 output in *Kalanchoe fedschenkoi*.[5] This shows a single peak in the 650- to 660-nm region.

Zimmer,[6] working with the petal movement rhythm of *Kalanchoe blossfeldiana*, shows that if the plants are entrained to a light/dark cycle and then transferred to continuous darkness, the effect of flashes of light on the free-running rhythm varies with the subjective phase. The subjective phase is the part of the normal light/dark cycle of which the displayed response is typical. If a flash of light is given during the subjective day, the rhythm is affected only slightly, if at all. If, however, the flash of light is given during the early part of the subjective night, the rhythm is delayed. The phase is set back abruptly. If the flash of light is given increasingly later in the subjective night, the delay of the phase increases until a certain point in time is reached when the flash of light causes an abrupt advancement of the phase. It would seem that light given early in the subjective night acts as a dusk signal entraining the rhythm for night-time activities, whereas at the end of the subjective night the light stimulation acts as a dawn signal advancing and entraining the rhythm for day-time activities.

The role of phytochrome in this system has been further elucidated by Simon *et al.*[7] These workers show that the leaf movement of *Samanea saman* is both initiated and maintained by short periods of R. Five-minute pulses of R would induce a phase shift in this system, but this effect could be readily reversed by immediate irradiation with FR. Although Bunning and Moser[8] show that the leaf movement in *Phaseolus multiflorus* may be entrained by exposing the pulvinus alone to R to reverse this effect, the lamina of the leaf needs to be exposed to FR, and this affects the activity of the pulvinus indirectly. Continuous FR shortens the cycle and R lengthens the cycle. Phytochrome may not be the only photoreceptor involved in regulating plant circadian rhythms. Satter *et al.*[9] show that a 2-h B treatment will induce a phase shift in *Samanea*, and they offer evidence that this response is not reversible by FR.

Circadian measurement of time

A model of circadian rhythms was proposed by Njus *et al.*[10] These workers suggested that self-sustained rhythms depend on a feedback arrangement between ion/membrane permeability and ion distribution. The photoreceptor for the rhythm was hypothesised to be membrane-bound and to function as an ion port. The insensitivity of rhythm to temperature changes was explained in terms of adaptation of membrane lipids. Phase changes were envisaged as a result of the photoreceptor altering the precise position of transport proteins.

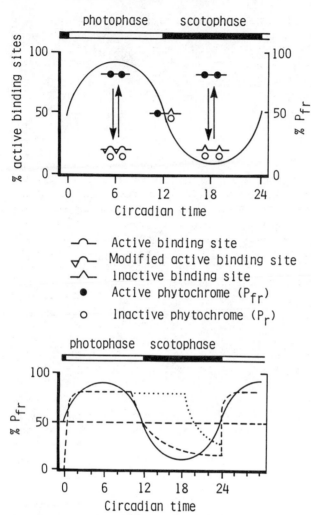

photophase   scotophase

⌐∿   Active binding site
∇∿   Modified active binding site
⌐∧   Inactive binding site
●   Active phytochrome (P_{fr})
○   Inactive phytochrome (P_r)

Figure 7.1   Heide's model of photoperiodic measurement of time (above) and an extension of the model to explain photoperiodic measurement of time in SDP (below) (redrawn from Heide, 1977).

Heide[11] has forwarded a model in which phytochrome has a pivotal role in a circadian rhythm timing system (see Figure 7.1). It is suggested that phytochrome is associated with structures which are responsible for membrane transport and thereby solute compartmentation. Phytochrome in the physiologically active form Pfr (a fact not universally agreed upon; see Smith 1983[12]) is bound to active binding sites during the photophase. R during this phase would reinforce the predominance of Pfr binding to active sites. FR during this phase converts Pfr to Pr, which does not bind to the active sites and can modify them. In the skotophase, as Pfr reverts to Pr, the active sites become inactive. These binding sites are therefore able to undergo circadian alterations which are linked to the photostationary state of phytochrome, and these changes alter the transport properties of the membrane. The changes in the receptor sites may also be enhanced by the changes in solute, pH, potential difference and other changes which occur with the transport activities associated with circadian time. Pfr is able to act as an allosteric effector to produce changes in its own receptor sites, which can account for re-phasing and entrainment.

Photoperiodism

Garner and Allard (1920)[13] were responsible for discovering and naming this phenomenon as a result of their work with the tobacco variety Maryland Mammoth and soy bean. Their research work has been described earlier (p. 114). Photoperiodism has been defined by Hillman[14] as a response to the timing of light and dark periods. In evidence discussed earlier (p. 118) it became apparent that it is the length of the dark period which is measured in photoperiodic responses, meaning that time is measured in an absolute rather then a relative fashion. Two types of mechanism might operate the photoperiodic timing in the flowering response. The first is often referred to as the hour glass mechanism and requires a series of sequential, unidirectional reactions to be initiated at the beginning of the dark period. Should each reaction proceed uninterrupted to completion, then flowering would be initiated. Much more evidence is available to support the second type of mechanism, which suggests that timing is carried out by the monitoring of a circadian rhythm. Cumming *et al.*[15] have shown that the flowering response of the SDP *Chenopodium rubrum* is governed by an endogenous rhythm. In their experiments, the results of which are shown in Figure 7.2, plants were grown under continuous incandescent light both before and after the experimental

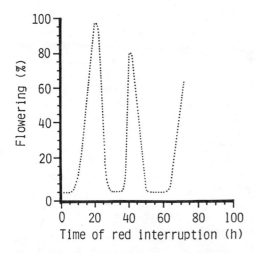

Figure 7.2 Demonstration of a rhythm in the sensitivity to R of the flowering response of *Chenopodium rubrum* during extended dark periods (Cumming, Hendricks and Borthwick, 1965).

period. Plants were given long dark periods up to 96 h, which were interrupted with R at different times. A rhythm is demonstrated in the responsivity to R. This has also been demonstrated for the SDP *Glycine max* and the LDP *Lolium temmulentum*. Thomas and Vince-Prue[16] have carried out similar investigations with the SDP *Pharbitis nil*. If seedlings are grown in the dark, exposed to WL for 24 h and then placed in the dark with R interruptions at various times, only one maxima of inhibition of flowering results. If the 24-h WL treatment is replaced by a pulse of R, then a maxima of inhibition appears coincident with the previous maxima, but in this case a weak rhythm develops. This rhythm can be reinforced if the plants are treated simultaneously with R and benzyladenine (a synthetic cytokinin).

Photoperiodic measurement of time

It is now generally accepted that circadian rhythms and photoperiodism are two manifestations of the same phenomenon. Two models have emerged which go some way to accommodate a huge accumulation of data. These are the external and internal coincidence models. In the external version, it is envisaged that there is a coincidence between the external light cycle and an internal rhythm. Light has two functions in this system. It sets the phase of rhythmic sensitivity to light, and it has another direct effect upon the photoperiodic phenomenon during the light-sensitive stage of the rhythm. In the internal coincidence model, light acts directly to entrain two separate rhythms, one with the 'light off' signal and one with the 'light on' signal. When a certain coincidence occurs between the phases of these two rhythms, a positive response ensues. The external coincidence model is a development of the Bunning hypothesis which suggests that plants oscillate between two 12-h phases, the photophil and the skotophil. Light in the photophil promotes and light in the skotophil, or dark period, inhibits photoperiodic phenomena. There is much more evidence to support the external coincidence model in plants than the internal model, which receives its support from animal systems. As was seen earlier (Chapter 6), evidence suggests that the measurement of time begins at the 'light off' signal and continues through the dark period, though the characteristics of this measurement do not agree with Pfr changes during the dark period. Thus dusk is thought to initiate or release an endogenous rhythm which determines the period over which light is inhibitory to flowering. When the dawn, or the 'light on' signal, occurs with the correct phase of the rhythm, then a positive response ensues.

The model proposed by Heide[11] to explain the measurement of circadian time was expanded to offer an explanation of photoperiodism, particularly in terms of the flowering response. Heide suggested that there were different optimal levels of Pfr in the photophase and skotophase which would promote flowering by maintaining or coinciding with a rhythm (see Figure 7.1). If a SDP were kept under cycles of 18 hL/6 hD, then the pattern of Pfr level throughout the cycle would be so far from the optimum required by the rhythm that flowering would be inhibited. On the other hand the behaviour of Pfr under 10 hL/14 hD would be sufficiently close to the optimal pattern to permit flowering. If the Pfr level deviates sufficiently from the optimum by an extension of the photophase, then a phase shift in the rhythm may result.

Heide's explanation for LDP is that the photophase is so long that it causes rephasing of the rhythm, and the difference between SDP and LDP is that in LDP this rephasing causes flowering.

Moonlight and photoperiodic measurement of time

Bunning and Moser[17] found that the lowest fluence rates of light which could influence circadian rhythms in *Glycine* corresponded with fluence rates supplied by moonlight. They suggested that it was possible that moonlight might disturb the measurement of time by plants. It was proposed that the circadian leaf movement in *Glycine*, *Arachis* and *Trifolium* could be of adaptive advantage because the orientation of the leaf in the relaxed position at night time reduces the incident radiation of the full moon to values of 5 to 20 per cent of those absorbed by a leaf surface in the near-horizontal day-time position. This received radiation is therefore predicted to fall below the threshold of response, and the circadian rhythms are able to proceed uninfluenced by moonlight. They suggest that in the SDP *Perilla ocymoides* and *Chenopodium amaranticolor* there is a specific photoperiodic phenomenon which counteracts the disturbing effects of moonlight. In these plants the effect of moonlight is to promote flowering instead of inhibiting it.

References for Chapter 7

1   Wilkins, M.B. 1959. An endogenous rhythm in the rate of carbon dioxide output of *Bryophyllum* I. Some preliminary experiments. *J. Exp. Bot.*, **10**, 377–390.

2   Salisbury, F.B. and Ross, C.W. 1978. *Plant Physiology*. Second Edition. California, Wadsworth, 304–315.

3   Bunning, E. and Stern, K. 1930. Uber die tagesperiodischen Bewegungender Primarblatter von *Phaseolus multiflorus*. II. Die Bewegungen bei Thermokonstanz. *Berichte der Deuschen Botanischen Gesellschaften*, **48**, 227–252.

4   Lorcher, L. 1957. Die Wirkung verschiedener Licht qualitaten auf die endogen Tagersrhythmik von *Phaseolus*. *Zeitschrift fur Botanik*, **46**, 209–241.

5   Wilkins, M.B. 1973. An endogenous circadian rhythm in the rate of carbon dioxide output of *Bryophyllum*. VI. Action spectrum for the induction of phase shifts by visible radiation. *J. Exp. Bot.*, **24**, 488–496.

6   Zimmer, K. 1962. Phasenvershiebung und andere Storlichtwirkungenauf die endogen tagesperiodischen Bluttenblattbewegungenvon *Kalanchoe blossfeldiana*. *Planta.*, **58**, 283–300.

7   Simon, E., Satter, R.L. and Galston, A.W. 1976. Circadian rhythmicity in excised *Samanea* pulvini. II. Resetting the clock by phytochrome conversion. *Plant Physiol.*, **58**, 421–425.

8   Bunning, E. and Moser, I. 1966. Respose-kurven bei der circadianen Rhythmic von *Phaseolus*. *Planta.*, **69**, 101–110.

9   Satter, R.L., Guggino, S.E., Longergan, T.A. and Galston, A.W. 1981. The effects of blue and red light on rhythmic leaflet movements in *Samanea* and *Albizzia*. *Plant Physiol.*, **67**, 965–968.

10 Njus, D., Sulzman, F.M. and Hastings, J.W. 1974. Membrane model for circadian clock. *Nature*, **248**, 116–120.

11 Heide, O.M. 1977. Photoperiodism in higher plants: an interaction of phytochrome and circadian rhythms. *Physiol. Plant*, **39**, 25–32.

12 Smith, H. 1983. Is Pfr the active form of phytochrome? *Phil. Trans. R. Soc. Lond.*, **303**, 443–452.

13 Garner, W.W. and Allard, H.A. 1920. Effect of the relative length of day and night and other factors of the environment on growth and reproduction in plants. *J. Agric. Res.*, **18**, 553–603.

14 Hillman, W.S. 1969. Photoperiodism and vernalisation. In: Wilkins, M.B. (ed.), *The Physiology of Plant Growth and Development*. London, McGraw-Hill, 559–598.

15 Cumming, B.G., Hendricks, S.B. and Borthwick, H.A. 1965. Rhythmic flowering responses and phytochrome changes in a selection of *Chenopodum rubrum. Can. J. Bot.*, **43**, 825–853.

16 Thomas, B., and Vince-Prue, D. 1984. Juvenility, photoperiodism and vernalization. In: Wilkins, M.B. (ed.), *Advanced Plant Physiology*. London, Pitman, 408–439.

17 Bunning, E. and Moser, I. 1969. Interference of moonlight with the photoperiodic measurement of time by plants and their adaptive reaction. *Proc. Nat. Acad. Sci. USA*, **62**, 1018–1022.

8
Special Situations

The development of the stereotypical higher plant has been dealt with in the preceding chapters. Wherever possible, exceptions to normal development have been dealt with alongside the generalisation. However, some reports in the literature do not fit easily into this format. Some of these are presented here. Some of the reports below concern special habitats and the role of light perception in these situations; others concern particular growth habits or life cycles which emphasise vegetative reproduction.

Aquatic

Aquatic angiosperms, particularly those of temperate climates, do not often persist in the vegetative state throughout the winter. In autumn, some form underground storage organs, while others produce turions as an overwintering organ. Turions vary in structure but are essentially bud-like organs containing a reduced shoot axis surrounded by young leaves. These leaves frequently contain copious quantities of storage material. Augsten *et al.* have investigated the role of light in the sprouting, often referred to as germination, of the turion of the duckweed *Spirodela polyrrhiza*.[1] They reported that in both light-grown and etiolated plants, the greatest response is found with R treatment and this effect is R/FR reversible with an escape time of 3 h, although the temperature of experimentation is not mentioned. Pulses of B, FR and green light also promote sprouting, particularly in etiolated plants, and the involvement of BAP cannot be eliminated.

From work carried out with *Myriophyllum verticillatum*[2] it was found that both temperature and photoperiod were involved in the formation of turions. Turion formation seems to be a LD/SD response since plants collected in the spring do not form turions easily in response to SD unless they are first exposed to LD. After LD treatment, turion formation is encouraged by low temperature (about 15°) and daylength of less than 16 h. Investigation of the influence of hormones on this situation suggests that turion induction in the autumn might be encouraged by an increase in endogenous ABA level with a concomitant decrease in cytokinin level. Before abscission, the growth of turions of *M. verticillatum* is promoted by LD but not by SD. After abscission, turions display a dormancy which can be fully broken by chilling at 4°. The dormancy can be relieved to a lesser degree by LD and at high temperature.[3]

Potamogeton crispus differs from other aquatic angiosperms in that not all populations develop turions; but in those which do, turions are produced in early summer rather than autumn.[4] The formation of these turions is promoted by LD greater than 16 h and high temperatures ($>16°$). Phytochrome is implicated in this response since a single, short night-break treatment with low-energy R will induce turion formation under SD. Under inductive conditions (LD and high temperature) the number of turions formed is reduced by low and increased by high fluence rates in a range of 50 to 400 μmol m^{-2} s^{-1}. A R : FR of 1, which is low in an aquatic environment (see p. 19), inhibits turion formation, whereas values above this cause promotion. These laboratory-demonstrated limitations in turion formation correlate well with observations of the plant in the lakes where the plant material was collected. Turion formation in this species takes place in the lakes of Scotland between June and August, when daylength is longer than 16 h and water temperatures are at their highest. Turion formation is largely limited to the upper 1.5 m of the water. Below this depth R : FR remains inductively high but PAR becomes inadequate.

Tropical rain forests

As a result of the high density of trees found in tropical rain forests and the ability of these species to compete for light, closed canopies develop. Tree species trying to establish themselves on the forest floor at low fluence rates exhibit three main strategies. Most species develop large seeds which are capable of germinating under low ϕ and establishing themselves as seedlings at low PAR. Although some species are capable of slow growth and develop through to maturity under the dense canopy conditions, the majority of species produce umbrophile seedlings, which depend upon a gap developing in the canopy above the seedling. The probability of such an event is about once in every hundred years[5] and, in consequence, the vast majority of seedlings in this ecological niche die off before such an opportunity arises. When a gap in the canopy does occur, an intense competition for light between developing trees begins. The third strategy is displayed by trees known as heliophile pioneers. These species produce large amounts of small seeds which remain dormant under canopy and only develop as a result of canopy break. The stimulation for germination in these species is light, and if the canopy closes after germination, these seedlings die. The light stimulation of the germination of seeds of two such tree species, *Cecropia obtusifolia* and *Piper auritum*, has been investigated.[6] *C. obtusifolia* occurs in large gaps in mature forest, growing at a rate of 3 m per year. It lives for less than 30 years but produces enormous numbers of small seed. *P. auritum* enjoys a similar habitat but reaches only 8 m in height and survives only 15 years, again producing large amounts of seed during this time. Germination experiments show that seeds of both these species are stimulated by high R : FR and inhibited and retarded by low ratios. Further investigation revealed that high levels of germination are evoked by long (2 h) periods of R daily for 3 successive days. This was interpreted as a mechanism by which the seed is able to ignore stimulatory effects of sunflecks and ensure germination in a more permanent area of sunlight.

Climbing mechanisms and shade adaptation

In a study of climbing woody vines, Darwin (1867)[7] arranged the mechanisms by which they climbed in what he thought to be the order of evolutionary advancement. He proposed that the most primitive method of climbing was used by the vines which produced adventitious roots from the stem to anchor the climber against an upright surface. Plants which had developed the ability to twine around the stems of other plants and elevate themselves were considered to be more advanced than the root-producing climbers. Other mechanisms such as the curling of contact-sensitive leaves and the use of leaf modifications such as tendrils, which in some plants are capable of producing an adhesive on contact, were considered further advancements. The rationale was that the production of roots required long-term contact with the climbing surface, whereas the faster reaction mechanisms of the other types were obviously selected to a higher degree. Furthermore, root-producing vines would be limited in their ability to cling to the more mobile upper parts of the canopy in which they were climbing and, in consequence, they would be unlikely to emerge into the high PAR region above the established canopy. It was also suggested that the twining mechanism had disadvantages in that young vines would be restricted by the availability of thin stems in an established tree canopy. Twining vines might be largely restricted to young trees and open canopies. Darwin also pointed out that leaf-climbing vines often retain the ability to twine but require less revolutions to maintain stability, thus investing less biomass in elevation. Tendril production was the most efficient in this respect, enabling vines to reach canopy tops with minimal biomass investment.

The twining vine, *Pueraria lobata*, has been reported to be shade-intolerant and is therefore poorly adapted to growing under forest canopy.[8] In a comparative study of the photosynthetic characteristics of eight vine species,[9] it was found that *P. lobata* had the highest light-compensation point ($43 \ \mu\text{mol m}^{-2}\text{s}^{-1}$), a low photosynthetic rate at low fluence rate ($0.5 \ \mu\text{mol m}^{-2}\text{s}^{-1}$ at a fluence of $50 \ \mu\text{mol m}^{-2}\text{s}^{-1}$) and the highest fluence requirement for 90 per cent maximal photosynthetic rate ($860 \ \mu\text{mol m}^{-2}\text{s}^{-1}$). It was, therefore, the most poorly adapted plant for surviving in canopy light of those studied. Another twining vine, *Lonicera japonica*, showed greater shade adaptation than *P. lobata*, but the species best adapted to photosynthesis in natural shade were the tendril-climbers. *Parthenocissus quinquefolia*, which climbs with the aid of adhesive tendrils, was the best adapted to the shade environment. It had a low compensation point of $20 \ \mu\text{mol m}^{-2}\text{s}^{-1}$, was able to photosynthesise at $3.5 \ \mu\text{mol m}^{-2}\text{s}^{-1}$ under the low fluence rate and showed light saturation at only $160 \ \mu\text{mol m}^{-2}\text{s}^{-1}$. Although equipped with the best climbing mechanism, this plant would not seem to be capable of exploiting the higher fluence rates found above the canopy. The root-climbers, *Hedera helix* and *Rhus radicans*, demonstrated the lowest maximum rate of photosynthesis and would seem to be able to survive by slow growth beneath the canopy. While not showing strong shade adaptations, they have neither the physiology nor the climbing mechanism to emerge and flourish above the canopy.

Light and buds

Kosinski and Giertych[10] were able to insert optical fibres into the apical domes in the buds of two conifers, *Pinus sylvestris* and *Picea abies*, in order to conduct natural light to the region of the female strobile. This treatment resulted in an increase in flowering. It was suggested that since the bud scales absorbed more R than FR, the Pfr level in the deeper tissues was low. By supplying unfiltered daylight, the Pfr level in the meristematic region was increased, leading to a promotion of flowering.

The precocious development of lateral buds is quite common in a number of species and is exaggerated in tomato, *Lycopersicon esculentum*, if the apical meristem is removed. Tucker[11] investigated the effect of end-of-day FR on the levels of auxin and ABA in the axillary bud of the tomato variety Craigella in specimens which were either intact or decapitated. The growth of buds of decapitated end-of-day-FR-treated plants remained inhibited, while those decapitated but receiving no end-of-day FR showed stimulation of outgrowth. Tucker suggested that end-of-day FR allows the synthesis of ABA in the mature leaves, which inhibits bud outgrowth. It is thought that IAA diffusing from the apex in non-decapitated plants serves to maintain the level of ABA in some way.

It may be that the influence of light on lateral buds varies slightly with plant habit. The opening of the lateral buds of the tree, *Salix viminalis*, shows a relationship with daylength. Whereas SD alone would not induce dormancy of buds, when used in conjunction with ABA, bud opening was prevented.[12]

Most commercially favoured potatoes produce tubers which have a short period of dormancy. If the buds are allowed to sprout in the darkness of a store, they produce long shoots which are easily knocked off during planting, thus delaying the emergence of the crop as other buds develop into shoots. McGee *et al.*[13] have shown that de-sprouting in most varieties after high-temperature storage ($22°$) not only leads to delayed emergence of the crop and delayed senescence but also to a decrease in yield (tuberisation). This cannot be entirely explained in terms of resetting the physiological age to zero at the time of planting. Light is used commercially to suppress bud growth of seed potatoes (*Solanum tuberosum*). This process, commonly known as chitting, results in the production of short, robust shoots. McGee *et al.*[14] quantified the light requirements of this process at different temperatures and showed that the effect of light had some of the same characteristics of the HIR as those reported for responses in etiolated tissue. The length of time that the tubers remained dormant was not affected by light, but the rate of growth once dormancy was broken was influenced by the daily period of exposure. Substantial irradiances were required over an extended period during which growth was linear with time at temperatures under $18°$.[15] The relationship with irradiance was log-linear and the dominant factor in determining the response was the total light energy received. Length of the photoperiod over which the energy is supplied has only a minor effect on the response. This response is closer to obeying the laws of reciprocity than is normally found in a HIR. McGee *et al.*[16] investigated the relationship of wavelength to suppression of bud growth in this species using 12 h photoperiods over 23 days. They found that inhibition of bud growth showed a maximum at 707 nm with a shoulder in the red. There was also a peak of inhibition in the B at more than 500 nm. In the work of Beggs *et al.*,[17] where the disappearance of the HIR was monitored in the inhibition of hypocotyl growth of *Sinapis alba* under certain light treatments (see p. 30), the maximum in the far-red region gives way to a maximum in the red region. This does not appear to happen with shoot growth in potatoes.

Tuberisation of potatoes

Since the early work of Garner and Allard (1923),[18] it has been accepted that LD suppresses tuberisation of potatoes. Their studies showed that tuberisation could be retarded by artificially extending the day beyond 14 to 15 h, and totally inhibited by daylengths of 18 h. Although these findings have been supported by others, recent work by Wheeler and Tibbitts[19] indicates that this effect should not be related to tuber yield in field situations. In this work the growth and tuberisation of five varieties were studied under continuous fluorescent illumination at two fluence rates (200 and 400 μmol m^{-2} s^{-1}), 12 h of the higher fluence extended by 12 h of 5 μmol m^{-2} s^{-1} of incandescent light and 12 h of high fluence followed by 12 h of dark. It was found that tuber development progressed well under 400 μmol m^{-2} s^{-1} and under 12 h of light per 12 h of dark, although tuber formation was suppressed under all other treatments. Continuous fluorescent illumination at both fluence rates caused stunting and leaf malformation in some of the varieties but not in others. No injury was apparent in other light treatments, but it was noted that by extending daylength with incandescent light, the plants showed significantly greater stem growth but gained less biomass than under other treatments. The discrepancy between the findings of these workers and others seems to be due to the fact that long irradiations favour shoot development over tuber formation, but on a comparative yield basis the extra photosynthetic activity predominates.

Bulbing

Onion (*Allium cepa*) is an obligate LDP and the bulbing response is photoperiod-dependent until maturity.[20] Lecari[21] shows that the addition of high-energy FR throughout the photoperiod shortens the length of the period required for bulbing. The response to added FR varies with the time at which it is given after the start of the photoperiod and may indicate the presence of a rhythmic response to the appropriate Pfr/Ptot. FR is the most effective in promoting bulbing when supplied in the middle of an 18 h photoperiod.[22] Approximately five times as much energy is required to bring about this 'all or nothing response' if the FR is supplied at the beginning or end of the photoperiod. An action spectrum for the photoperiodic induction of bulbing indicates a single action maximum at 714 nm.[23] The fluence rate dependency of the response to FR is characteristic of the HIR.

Tap root

The Sweet Clovers, *Melilotus alba* and *M. officinalis*, are biennials at southerly latitudes. In the first year the seed germinates in the spring and the plants grow slowly to a height of about 50 cm. At the end of the first year the aerial parts die back and the plant overwinters as an enlarged tap root. In the second year adventitious buds develop at the top of the tap root. The plant, which grows to a height of about 150 cm, goes on to flower and produce seed. Kasperbauer et al.[24] showed that flowering in the second season was enhanced by vernalisation. They also reported that tap root formation depended on SD treatment. In Alaska, with the prolonged arctic summer days, these species can be induced to flower in the first year and do not form tap roots.

Tendrils

Tendrils are long, thin, hair-like organs which can be modified stems, leaves or flower peduncles. They are found in plants which secure themselves to supporting structures to maintain an erect habit. Jaffe and Galston[25,26] show that pea tendrils, which are leaflet modifications, circumnutate until they make contact with a structure. Circumnutation then ceases and coiling begins. Tendrils which had been held in the dark overnight were found to coil less well then tendrils held in the light. Jaffe[27] was able to show that if tendrils were held in the dark for several days they would not coil when stimulated until they had received further illumination. The tendril would coil if the light stimulus was supplied within 1 to 2 h of the physical stimulation, after which further physical stimulation was required to cause coiling. Shotwell and Jaffe[28] provide an action spectrum of the light stimulation effect which shows a shoulder at 480 nm and a clear maximum at 440 nm.

CAM metabolism

Plants which grow in arid regions frequently display a modification known as crassulacean acid metabolism (CAM). Named after the group of plants in which it was first reported, CAM metabolism is now known in about 25 to 30 families and is common among the members of these families which grow in habitats where water retention is of prime importance. These plants have developed the ability to regulate their stomata so that they are closed during the high-temperature period of daylight hours but open to allow gaseous exchange during the night period, when water loss is minimal. The CO_2 from the environment is only available during the period when the light reactions of photosynthesis are not functioning. In consequence, a mechanism has evolved which fixes the CO_2 available at night time into malate, which is stored in the vacuoles of the cells of the photosynthetic tissue until light is available to provide the reductive energy required to fix carbon via the Calvin or C3 cycle. Under these conditions, the malate is transported out of the vacuole and broken down and the carbon made available for fixation.

Many CAM plants are obligate but others are facultative. Facultative CAM plants may be switched from C3 mode of photosynthesis to CAM by various environmental stimuli such as water stress or daylength. Water stress is known to induce CAM directly in *Sedum spectabile*[29] and indirectly in *Mesembry-anthemum crystellinium* via salinity.[30] In the former species, it was found that under well-watered conditions plants remained C3 irrespective of daylength. Transfer to drought conditions caused a shift to CAM within 12 h, as measured by CO_2-exchange patterns. As soon as the plants were rewatered they reverted to C3. It was also shown that appropriate LD would enhance the shift to CAM. Probably the best investigated photoperiodic induction of CAM is that of *Kalanchoe blossfeldiana* var. Tom Thumb.[31,32] In this plant, CAM was irreversibly induced by SD, and phytochrome was shown to be involved in this response. The synthesis of PEPcarboxylase, the enzyme which fixes CO_2 when CAM is induced, was shown to be controlled by phytochrome[33] and the crucial step in the induction of CAM. It has been suggested by Queiroz and Brulfert[34] that SD not only induces CAM but also activates a sequence of genetically programmed changes which prepares the plant for oncoming drought. The proof of this argument depends upon finding that C3-performing plants are less adapted, other than in their stomatal cycles, to drought conditions. Brulfert *et al.*[35] compare LD-grown C3-performing Tom Thumb to SD-grown CAM-performing plants. When both sets of plants were subjected to drought, CAM was induced under LD after 48 h. The CAM metabolism in the SD-grown plants was increased without lag phase, but otherwise the CAM induced by SD and the CAM induced by water stress were comparable, and these workers suggest that this indicates that different messengers, triggered by either phytochrome or water stress, act via the same target as a first step in the induction pathway.

Frost tolerance

It has been known for some time that the frost-hardiness of woody dicotyledonous plants may be influenced by photoperiod.[36, 37] It was suggested that cold acclimation, like many other photoperiodic phenomena, may be influenced by phytochrome.[38] Indeed, Williams *et al.*[39] show that if *Cornus stolonifera* and *Weigela florida* were given SD and the dark period was interrupted with R, then the SD effect on cold-hardiness was nullified. The winter rape seedling, *Brassica napus*, has been studied in this connection.[40] It was found that WL, B and R all increased frost-hardiness in 3-day-old seedlings and inhibited elongation of the hypocotyls, although the most effective treatment was R. Both effects were reversed either by FR or by the application of GA_3. Frost damage is largely caused by water crystallising in the cells and, since light-treated plants have short hypocotyls, they contained less water than etiolated plants, but there may be a more direct effect since R-treated plants lose more water during the freezing process than do B-treated plants.

The grass *Lolium perenne* has been shown to require hardening periods of low temperatures above freezing to induce frost tolerance.[41] It has been shown by others[42] that light conditions during the pre-hardening and the hardening period are also important. Apparently a reduction in total fluence is capable of reducing hardening whether the light is given during the pre-hardening or hardening period. However, a reduction of photoperiod or fluence rate has different effects at different phases. The greatest reduction of cold tolerance that could be achieved during the pre-hardening (20°) period was due to LD treatment in combination with low fluence rates, whereas during the hardening phase the greatest reduction was achieved with a combination of high fluence rate and short days.

Winter cereals, when surrounded by ice at temperatures just below freezing, form an anaerobic environment and have been shown to accumulate toxic substances such as ethanol, which together with CO_2 can slowly reach toxic levels.[43,44] Since previous studies of ice encasement had been made in the dark, Andrews[45] has investigated the effect of light treatment during this period, since the effect of light on photosynthesis might significantly alter this situation. He found that in a number of winter wheat varieties, a winter barley and a winter rye, exposure to $100 \, \mu mol \, m^{-2} s^{-1}$ during ice encasement at $-1°C$ markedly increased survival rate. The most significant metabolic changes after encasement in the light are that less ethanol but more CO_2 is found in plant crowns. It was suggested that low-level photosynthesis was responsible for these changes, supplying the cells with greater levels of energy, which aided cell maintenance during ice encasement.

References for Chapter 8

1   Augsten, H., Kunz, E. and Appenroth, K-J. 1988. Photophysiology of turion germination in *Spirodela polyrrhiza*. I. Phytochrome-mediated responses of light- and dark-grown turions. *J. Plant Physiol.*, **132**, 90–93.

2   Weber, J.A. and Nooden, L.D. 1976. Environmental and hormonal control of turion formation in *Myriophyllum verticillatum*. *Plant cell Physiol.*, **17**, 721–731.

3   Weber, J.A. and Nooden, L.D. 1976. Environmental and hormonal control of turion germination in *Myriophyllum verticillatum*. *Am. J. Bot.*, **63**, 936–944.

4   Chambers, R.A., Spence, D.H.N. and Weeks, D.C. 1985. Photocontrol of turion formation by *Potamageton crispus* in the laboratory and in natural water. *New Phytol.*, **99**, 183–194.

5   Whitemore, T.C. 1975. *Tropical Rainforests of the Far-East*. Oxford, Clarendon Press, 278.

6   Vazquez-Yanes, C. and Smith, H. 1983. Phytochrome control of seed germination in the tropical rainforest pioneer trees *Cecropia obtusifolia* and *Piper auritum* and its ecological significance. *New Phytol.*, **92**, 477–486.

7   Darwin, C. 1867. On the movements and habits of climbing plants. *J. Linn. Soc. London, Bot.*, **9**, 1–118.

8   Wechsler, N.R. Growth and physiological characteristics of kudzu, *Pueraria lobata* in relation to its competitive success. M.S. Thesis University of Georgia, Athens GA.

9   Carter, G.A. and Teramura, A.H. 1988. Vine photosynthesis and relationships to climbing mechanics in a forest understorey. *Amer. J. Bot.*, **75**, 1011–1018.

10 Kosinski, G. and Giertych, M. 1982. Light conditions inside developing buds affect floral induction. *Planta.*, **155**, 93–94.

11 Tucker, D.J. 1977. The effects of far-red light on lateral bud outgrowth in decapitated tomato plants and the associated changes in the levels of auxin and abscissic acid. *Plant Sci. Lett.*, **8**, 339–344.

12 Barros, R.S. and Neill, S.J. 1986. Periodicity of response to abscisic acid in lateral buds of willow (*Salix viminalis*). *Planta.*, **168**, 530–535.

13 McGee, E., Booth, R.H., Jarvis, M.C. and Duncan, H.J. 1988. The inhibition of potato growth by light. III. Effects on subsequent growth in the field. *Ann. Appl. Biol.*, **113**, 149–157.

14 McGee, E., Booth, R.H., Jarvis, M.C. and Duncan, H.J. 1987a. The inhibition of potato sprout growth by light. I. Effects of light on dormancy and subsequent sprout growth. *Ann. Appl. Biol.*, **110**, 399–404.

15 McGee, E., Jarvis, M.C. and Duncan, H.J. 1986. The relationship between temperature and sprout growth in stored seed potatoes. *Potato Research.*, **29**, 521–524.

16 McGee, E., Jarvis, M.C. and Duncan, H.J. 1987. Wavelength dependence of suppression of potato sprout growth by light. *Plant, Cell and Environ.*, **10**, 655–660.

17 Beggs, C.J., Holmes, M.C., Jabben, M. and Schafer, E. 1980. Action spectra for the inhibition of hypocotyl growth by continuous irradiation in light and dark-grown *Sinapis alba* seedlings. *Plant Physiol.*, **66**, 615–618.

18 Garner, W.W. and Allard, H.A. 1923. Further studies in photoperiodism, the response of the plant to the relative length of day and night. *J. Agric. Res.*, **23**, 871–920.

19 Wheeler, R.M. and Tibbitts, T.W. 1986. Growth and tuberisation of potato (*Solanum tuberosum*) under continuous light. *Plant Physiol.*, **80**, 801–804.

20 Magruder, R. and Allard, H.A. 1937. Bulb formation in some American and European varieties of onions as affected by length of day. *J. Agric. Res.*, **54**, 719–752.

21 Lecari, B. 1982. The promoting effect of far-red light on bulb formation in the long day plant *Allium cepa*. *Plant Sci. Lett.*, **27**, 243–254.

22 Lecari, B. and Deitzer, G.F. 1987. Time-dependent effectiveness of far-red light on the photoperiodic induction of bulb formation in *Allium cepa*. *Photochem. Photobiol.*, **45**, 831–835.

23 Lecari, B. 1983. Action spectrum for the photoperiodic induction of bulb formation in *Allium cepa*. *Photochem. Photobiol.*, **38**, 219–222.

24 Kasperbauer, M.J., Gardner, F.P. and Loomis, W.E. 1962. Interaction of photoperiod and vernalisation in flowering of sweet clover (*Melilotus*). *Plant Physiol.*, **37**, 165–170.

25 Jaffe, M.J. and Galston, A.W. 1966a. Energetics of contraction and coiling in pea tendrils. *Plant Physiol.*, **41**, suppl-43.

26 Jaffe, M.J. and Galston, A.W. 1966b. Physiological studies on pea tendrils. II. The role of light and ATP in contact coiling. *Plant Physiol.*, **41**, 1152–1158.

27 Jaffe, M.J. 1977. Experimental separation of sensory and motor functions in pea tendrils. *Science*, **195**, 191–192.

28 Shotwell, M. and Jaffe, M.J. 1979. Physiological studies on pea tendrils. X. Characterisation of the light activation effect on contact coiling as a blue light trigger. *Photochem. Photobiol.*, **29**, 1153–1156.

29 Brulfert, J.M., Kluge, S., Guclu, S. and Queiroz, O. 1988a. Combined effects of drought, daylength and photoperiod on rapid shifts in the photosynthetic pathways of *Sedum spectabile*, a CAM species. *Plant Physiol. and Biochem.*, **26**, 7–16.

30 Winter, K. and Luttge, U. C3 Photosynthese und Crassulaceen-Saurestoffwechsel bei *Mesembryanthemum crystillinium*. *Ber. Dtsch. Bot. Ges.*, **92**, 117–132.

31 Gregory, F.G., Spear, J. and Thiaman, K.V. 1954. The interrelation of CO_2 metabolism and photoperiodism in *Kalanchoe*. *Plant Physiol.*, **29**, 220–229.

32 Queiroz, O. 1968. Sur le metabolisme acide des Crassulacees. III. Variations d'activité enzymatique sous l'action du photoperiodisme et du thermoperiodisme. *Physiol. Veg.*, **6**, 117–136.

33 Brulfert, J., Vidal, J., Keryer, M., Thomas, P., Gadal, P. and Queiroz, O. 1985. Phytochrome control of PEP carboxylase synthesis and specific RNA level during photoperiodic induction in a CAM plant during greening in a C4 plant. *Physiol. Veg.*, **23**, 921–928.

34 Queiroz, O. and Brulfert, J. 1982. Photoperiod-controlled induction and enhancement of seasonal adaptation to drought. In: Ting, I.P. and Gibbs, M. (eds.), *Crassulacean Acid Metabolism*. Baltimore, Waverly, 208–230.

35 Brulfert, J., Kluge, M., Guclu, S. and Queiroz, O. 1988b. Interaction of

photoperiod and drought as CAM inducing factors in *Kalanchoe blossfeldiana* cv. Tom Thumb. *J. Plant Physiol.*, **133**, 222–227.

36 Irving, R.M. and Lanphear, F.O. 1968. Regulation of cold hardiness in *Acer negundo*. *Plant Physiol.*, **43**, 9–13.

37 Fuchigami. L.H., Weiser, C.J. and Evert, D.R. 1971. Induction of cold acclimation in *Cornus stolonifera*. *Plant Physiol.*, **47**, 98–107.

38 Weiser, C.J. 1968. Endogenous control of cold acclimation in woody plants. *Cryobiol.*, **4**, 277.

39 Williams, B.J., Pellet, N.E. and Klein, R.M. 1972. Phytochrome control of growth cessation and initiation of cold acclimation in selected woody plants. *Plant Physiol.*, **50**, 262–265.

40 Kacperaka-Palacaz, Debska, Z. and Jakubowska, A. 1975. The phytochrome involvement in the frost hardening of rape seedlings. *Bot. Gaz.*, **136**, 137–140.

41 Lorenzetti, F., Tyler, B.F., Cooper, J.P. and Breese, E.L. 1971. Cold tolerance and winter hardiness in *Lolium perrene*. I. Development of screening techniques for cold tolerance and survey of geographical variations. *J. Agric. Sci.*, **76**, 199–209.

42 Lawrence, T., Cooper, J.P. and Breese, E.L. 1973. Cold tolerance and winter hardiness in *Lolium perrene*. II. Influence of light and temperature during growth and hardening. *J. Agric. Sci.*, **80**, 341–348.

43 Andrews, C.J. 1977. The accumulation of ethanol in ice-encased winter cereals. *Crop Sci.*, **17**, 157–161.

44 Andrews, C.J. and Pomeroy, M.K. 1979. Toxicity of anaerobic metabolites accumulating in winter wheat seedlings during ice-encasement. *Plant Physiol.*, **64**, 120–125.

45 Andrews, C.J. 1988. The increase in survival of winter cereal seedlings due to light exposure during ice-encasement. *Can. J. Bot.*, **66**, 409–413.

Index

Index